Rational Pharmacotherapy of the Inner Ear

Advances in Oto-Rhino-Laryngology

Vol. 59

Series Editor

W. Arnold *Munich*

KARGER

Rational Pharmacotherapy of the Inner Ear

Volume Editors

D. Felix *Berne*
E. Oestreicher *Munich*

44 figures, 3 in color, and 9 tables, 2002

KARGER

Basel · Freiburg · Paris · London · New York ·
New Delhi · Bangkok · Singapore · Tokyo · Sydney

Prof Dr. D. Felix
Division of Neurobiology
University of Berne
Baltzerstrasse 6
CH–3012 Berne (Switzerland)

Dr. E. Oestreicher
ENT-Department
Technical University of Munich
Ismaninger Strasse 22
D–81675 Munich (Germany)

Bibliographic Indices. This publication is listed in bibliographic services, including Current Contents® and Index Medicus.

Drug Dosage. The authors and the publisher have exerted every effort to ensure that drug selection and dosage set forth in this text are in accord with current recommendations and practice at the time of publication. However, in view of ongoing research, changes in government regulations, and the constant flow of information relating to drug therapy and drug reactions, the reader is urged to check the package insert for each drug for any change in indications and dosage and for added warnings and precautions. This is particularly important when the recommended agent is a new and/or infrequently employed drug.

www.karger.com
Printed in Switzerland on acid-free paper by Reinhardt Druck, Basel
ISSN 0065–3071
ISBN 3–8055–7279–4

Contents

Vascular System – Ion Homeostasis

Immunology – Cell Cycle

Ototoxicity, Noise and Aging

Preface

This 59th volume of the series *Advances in Oto Rhino-Laryngology* is dedicated to the evolving area of inner ear pharmacology. Sudden hearing loss, progressive hearing loss, presbycusis, noise trauma, Ménière's disease, ototoxicity and tinnitus are examples of inner ear disorders that often cannot be treated sufficiently. Although a wide range of interesting new concepts in basic research on inner ear physiology, pathophysiology and therapy have been published over the last years, only few have been adapted to the medical management of inner ear diseases. This might be due to a still deficient exchange between experts in basic research and clinicians in the field of otology and inner ear diseases. The present volume is a compilation of reviews and reports dealing with all aspects of inner ear research (physiology, anatomy, neurotransmission, vascular system, differentiation, regeneration and immunology) as well as reports on experimental inner ear therapy and clinical trials on new inner ear treatment strategies. It provides an overview on recent knowledge on inner ear pharmacology by some of the experts in this field.

Most of the included reports and reviews were presented at the Second International Symposium on Inner Ear Pharmacology on 'Rational Pharmacotherapy of the Inner Ear – from Basic Research to Clinics', held in Wildbad Kreuth, Germany, from October 5 to 8, 2000. Inspired by the First Symposium on the Pharmacology of the Inner Ear, held in Montpellier in 1994, the objective of the second meeting was to bring together a representative faculty of experts in basic research and clinicians interested in otology and inner ear diseases. The fruitful discussions of these two groups should stimulate the design of future treatment strategies of inner ear diseases based on recent scientific knowledge on the pharmacology of the inner ear.

It is the hope of the editors that this volume will be further step towards the development of new treatments of inner ear diseases as a result of the collaboration between basic scientists and clinicians. In this regard the volume should be interesting to both scientists and clinicians and inspire future research on the inner ear.

Munich, Berne, October 2001

Elmar Oestreicher
ENT Department
Technical University of Munich
Dominik Felix
Department of Neurobiology
University of Berne

Felix D, Oestreicher E (eds): Rational Pharmacotherapy of the Inner Ear.
Adv Otorhinolaryngol. Basel, Karger, 2002, vol 59, pp 1–10

Anatomical Differences in the Peripheral Auditory System of Mammals and Man

A Mini Review

Heidi Felix

ENT Department, University Hospital of Zürich, Switzerland

Abstract

The major anatomical differences among animal models and man are briefly reviewed. Differences are described in the length and width of the basilar membrane, the number of inner (IHCs) and outer hair cells (OHCs), and the length of cilia on both cell types. Significant differences in the innervation pattern of the IHCs among these species include the number of afferent nerve terminals per IHC, the degree of branching of afferent fibers and the number of synapses per afferent nerve terminal. At the OHCs, the number of afferent and efferent nerve terminals, the presence or absence of presynaptic bodies, reciprocal synapses and the presence of dendrodendritic synapses in the outer spiral bundles may have important physiological functions. In the cochlear nerve, significant differences are described in the number of spiral ganglion cells (SGCs) and cochlear nerve fibers. Furthermore the percentage of myelinated SGCs and the presence of synapses on SGCs varies enormously.

The purpose of this mini review is to summarize data concerning anatomical differences of the cochlea and the cochlear nerve among the most common mammalian models for the study of human hearing. Despite many similarities in the cochlea and the cochlear neurons of these species, there are a number of anatomical differences which may have physiological implications. It is of importance to recognize these differences, because the applicability of mammalian models for studying human auditory function and dysfunction depends at least in part on the similarity of the anatomy of the auditory system across these species.

Table 1. Cochlea

	Rat	Guinea pig	Cat	Rhesus monkey	Man
Number of cochlear turns	2.5 [1]	4.5 [2]	3.5 [1]	3 (Felix, unpubl.)	2.5 [2]
Basilar membrane length, mm	5 [1]	19–21 [2]	23 [2]	27–29 (Felix, unpubl.)	28–40 [3]
Basilar membrane width, μm	B 100 M 200 A 250 [7]	B 62 M 197 A 209 [6]	M 350 [4]		B 104 M 336 A 504 [5]

B = Basal turn; M = Middle turn; A = Apical turn.

Cochlea

In almost all mammalians and in humans, the cochlea (table 1) is divided into three compartments, the scala tympani, media and vestibuli spiraling around the modiolus. The number of cochlear turns varies from species to species [1], and the length of the basilar membrane differs widely among mammalian species but also within species [2, 3]. The width of the basilar membrane increases progressively from the base to the apex in all investigated species [4–7]. The different lengths and widths of the basilar membrane presumably reflect differences in physiological frequency ranges.

Organ of Corti

The organ of Corti with the two types of receptor cells (inner, IHCs, and outer hair cells, OHCs) is located on the basilar membrane forming the floor of the cochlear duct. The organ of Corti is innervated by afferent, efferent and autonomic neurons.

Inner Hair Cells. Corresponding to the different lengths of the basilar membrane the number of hair cells varies among mammalian species [6–10] (table 2). In all species studied, the length of cilia at the apical pole of the IHCs increases from base to apex [7, 11, 12].

According to Spoendlin [13], about 95% of the afferent fibers innervate the IHC system in the cat. In guinea pigs [14] and man [15], about 90% of afferent nerve fibers form connections with IHCs. At the neural pole of the IHCs, afferent nerve endings are less numerous in humans than in cats [16–19]. In the cat,

Table 2. Inner hair cells

	Rat	Guinea pig	Cat	Rhesus monkey	Man
Number	960 [10]	2,400 [9]	2,600 [4]	2,400–2,600 (Felix, unpubl.)	2,800–4,400 [8]
Number of cilia/IHC	B 32 M 28 A 41 [11]	M 34–40 [12]			B 41–44 M 64 [5]
Afferent innervation, %		90 [14]	95 [13]		90 [15]
Afferent nerve terminals/IHC			20 [16] 26 [17]		B 6/9 M 12/11 A 9/11 [18, 19]
Synapses/terminal			1 [16, 17]		B 15 M 18 A 16 [20]
Presynaptic body		yes [21]	yes [16, 17]	yes [22]	yes [20]
Branching of afferent fibers		yes [23]	rare [16] no [17] yes [24]	yes [22]	yes (to 3 cells) [20]
Efferent terminals on IHC		rare [16]	no [17]		no [20]
Axodendritic efferents		yes [25]	yes [26]	yes [26]	yes [20]

For abbreviations, see table 1.

there is only one synapse between each radial fiber and IHC [16, 17], whereas multiple synapses – up to 18 in the middle turn per terminal – are present in humans [20]. Presynaptic bodies have been found in all species [16, 17, 20–22]. Branching of radial fibers seems to be common in man [20] as well as in the guinea pig [23] and monkey [22]. In the cat, the findings about branching are controversial between the different authors [16, 17, 24]. Efferent nerve endings on IHCs are rare in the guinea pig [25] and lacking in cats and man [17, 20], but in all species there is evidence for axodendritic terminations of efferent fibers near the base of IHCs [20, 25, 26].

Table 3. Outer hair cells

	Rat	Guinea pig	Cat	Rhesus monkey	Man
Number	3,470 [10]	8,000 [9]	9,900 [4]	9,000–10,000 (Felix, unpubl.)	11,200–16,000 [8]
Number of cilia/OHC	B 75 M 96 A 62 [11]	B about 115 M about 115 [12]	B 97 [28]	B 121 [28]	B 120–148 A 46–80 [27]
Afferent nerve terminals/OHC		B 4 A 15 [21] A 8 [29]	B 6–10 A 8 [16]		7–10 [7] B 4 M 5 A 8 [18]
Presynaptic body		yes [22]	no [24]	yes [30]	yes (35%) [31]
Reciprocal synapses		no [29]	pathological only [33]		in about 56% [32]
OHC innervated by one SGC		M 6–20 [34]	B 6–8 A 20 [35]		
Efferent terminals/OHC	present [36]	A 7 [29] A 10–16 [21]	present [16]	present [22]	present [18]
Axodendritic efferents		yes [21]	yes [16]	yes [22]	yes [31]
Dendro-dendritic synapses				yes [37]	yes [32]

For abbreviations, see table 1.

Outer Hair Cells. The number of OHCs shows a rather wide variation across mammalian species [6, 8–10] (table 3). The OHCs have a greater number of cilia per hair cell [11, 12, 27, 28] compared to IHCs, and the length of cilia [6, 11] is remarkably constant across species and similar to IHC dimensions.

In the OHCs, the afferent endings [7, 16, 18, 21, 29] are fewer than in IHCs. Unlike that in IHCs, the innervation of OHCs is convergent. Although not all synapses have a presynaptic body [22, 24, 30], the synaptic morphology is consistent with a chemical synapse [31]. Reciprocal synapses at the base of OHCs have been described in humans [32]. This suggests both hair-cell-to-neuron and neuron-to-hair-cell synaptic activity within one neural ending. So far reciprocal synapses have not been identified in the organ of Corti of other mammals [29]

Table 4. Cochlear neurons

	Rat	Guinea pig	Cat	Rhesus monkey	Man
Number of SGCs	15,800 [10]		44,298–57,494 [38]		23,200–39,100 [39] 29,800–38,350 [40]
Number of axons		24,011 [41]	51,755 [41]	31,247 [41]	21,600–32,700 [19] 31,400 [42]
Myelination of SGCs, %	90–93 [48]	90 [50]	95 [43, 44]	85 [45]	<5 [46, 47]

except as a pathological change following section of the olivocochlear bundle [33]. In the guinea pig, Smith [34] found that in the middle turn 1 spiral ganglion cell (SGC) innervates 6–20 OHCs whereas, according to Simmons [35], in the basal turn of the cat, each type 2 SGC innervates 6–8 OHCs and in the apical turn up to 20 OHCs. In contrast to the IHCs there are efferent nerve endings directly on OHCs across all mammalian species investigated [16, 18, 21, 22, 29, 36]. There are also synaptic contacts between efferent and afferent fibers [16, 21, 22, 31] as well as between afferent fibers in the monkey and man [32, 37].

Cochlear Neurons. The differences in the first-order neurons are even more pronounced (table 4). A wide variation among various mammalian species in the number of SGCs [10, 38, 39] and central nerve processes exists [19, 40, 41]. Based on the diameter of the cell body there appear to be two classes of ganglion cells in cats [43, 44], monkeys [45] and man [46, 47]. Such a bimodal distribution is not apparent in the guinea pig [44]. Recent morphometric investigations provided evidence of 3 types of ganglion cells in the human spiral ganglion: large, intermediate and small, varying from each other significantly on the basis of cell area [50]. The authors suggested that based on the morphological findings the large and intermediate ganglion cells were subtypes of the classic type 1 SGC, whereas the small ganglion cell was consistent with the classically described type 2 SGC. In the rat [48] and guinea pig [49], over 90% of the SGCs are myelinated (fig. 1). In the cat [43, 44], about 95% of the cells are large in size (type 1) and have myelinated cell bodies whereas in the monkey only about 85% of the cells are large and myelinated [45]. The type 2 cells are small in size and unmyelinated [43–45]. In contrast to all mammalian species so far investigated, myelination of SGCs in humans is rare (fig. 2), and a variety of cell shapes have been described [46, 47, 51]. Furthermore synapses on human small (type 2) SGCs [52] as well as on large SGCs (type 1) have been described

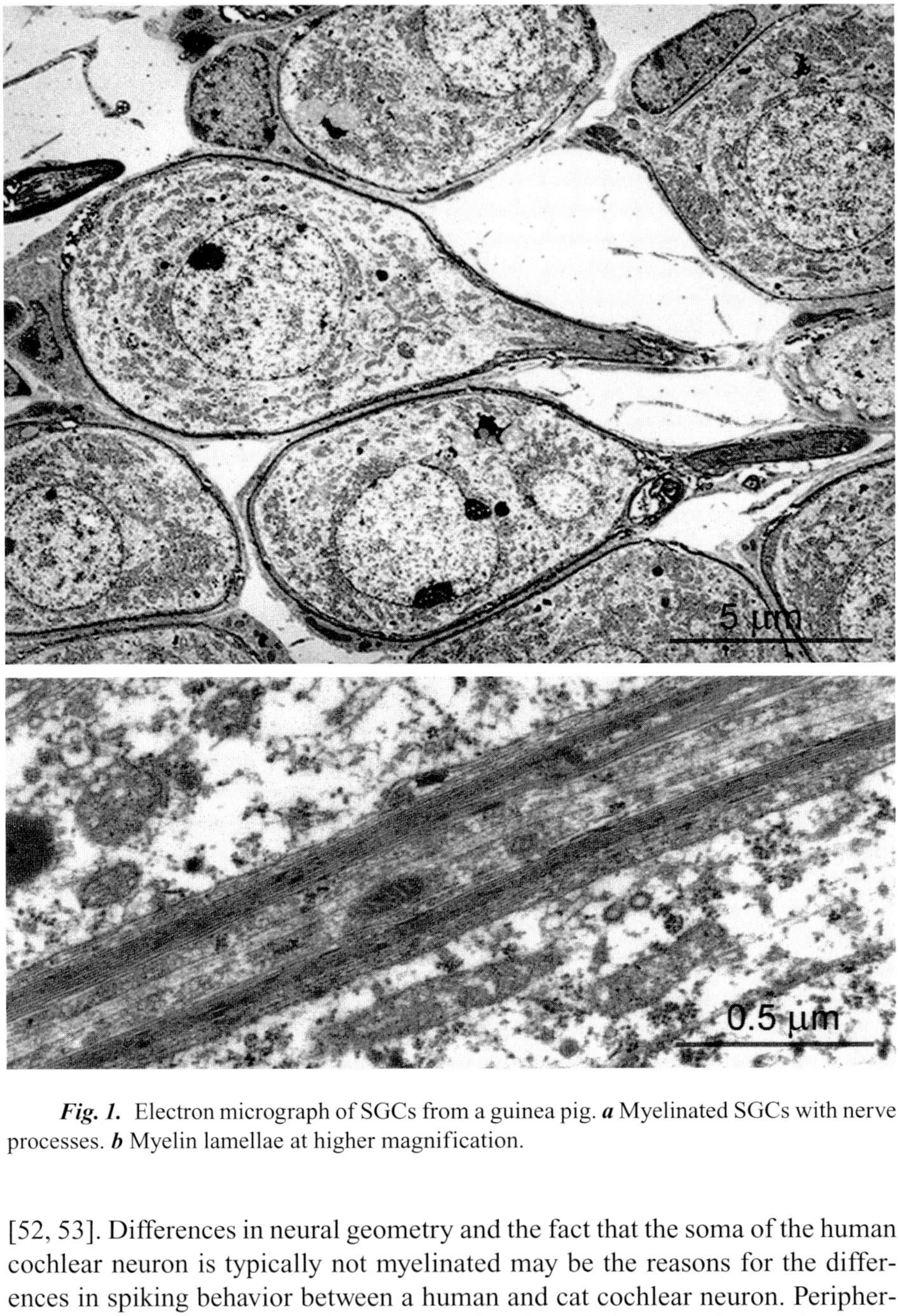

a

b

Fig. 1. Electron micrograph of SGCs from a guinea pig. ***a*** Myelinated SGCs with nerve processes. ***b*** Myelin lamellae at higher magnification.

[52, 53]. Differences in neural geometry and the fact that the soma of the human cochlear neuron is typically not myelinated may be the reasons for the differences in spiking behavior between a human and cat cochlear neuron. Peripherally and centrally evoked spikes arrive with a time difference of about 400 µs in man and 200 µs in the cat [54].

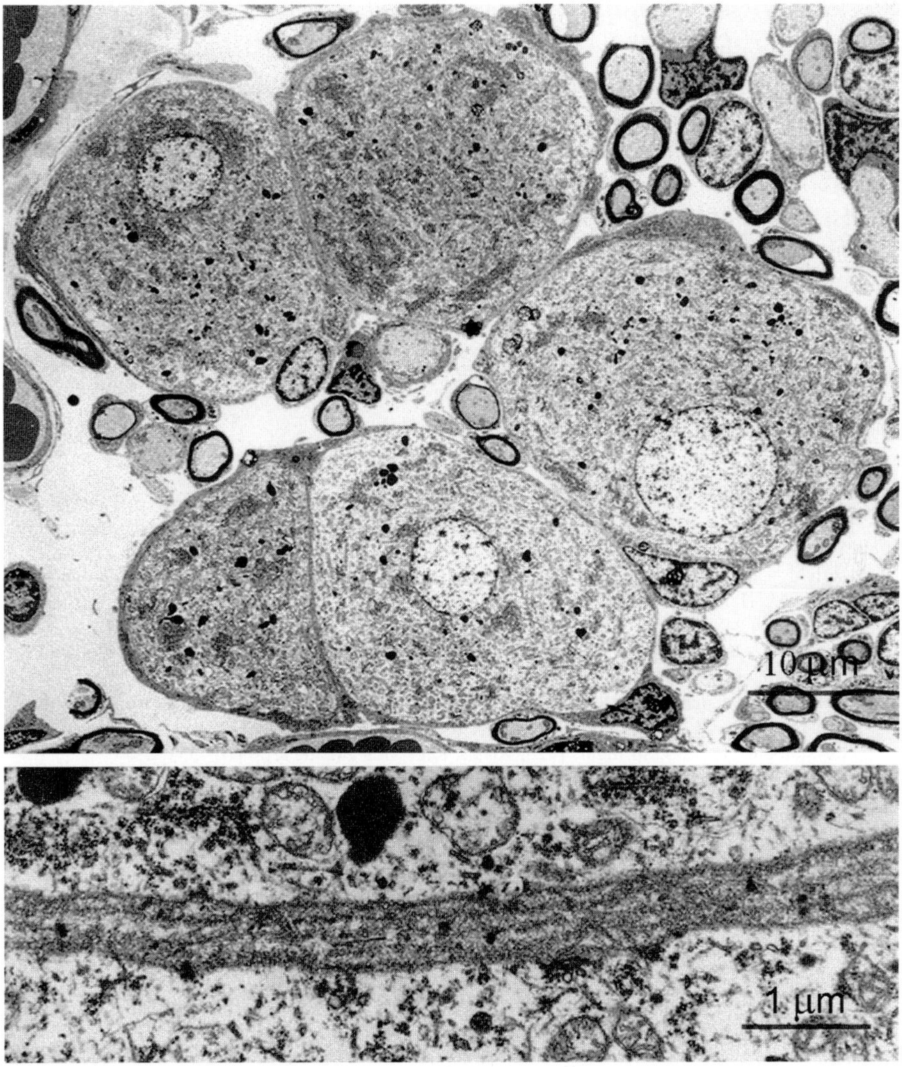

Fig. 2. Electron micrograph of human SGCs. ***a*** Nonmyelinated SGCs. ***b*** Cell membrane of SGC without myelin lamellae but flanked by satellite cells.

Conclusion

Despite many similarities both in the cochlea and the cochlear neurons of these mammalian species, there are a number of differences which may have important physiological implications.

In the cochlea, the major differences among these species are the length and width of the basilar membrane, the number of IHCs and OHCs and the length of cilia on both IHCs and OHCs.

Furthermore the innervation patterns of the human organ of Corti differ significantly from most animal models. The anatomical differences, such as innervation density of hair cells, the presence of reciprocal synapses at the base of OHCs, lack of myelination in the great majority of SGCs and the presence of synapses on SGCs, are likely to reflect important physiological differences that need to be considered in modelling the function of the normal human inner ear.

References

1 Igarashi M, Mahon RC Jr, Konishi S: Comparative measurements of cochlear apparatus. J Speech Hear Res 1968;11:229–235.

2 Spoendlin H: Strukturelle Organisation des Innenohres; in Naumann HH, Helms J, Herberhold C, Kastenbauer E (eds): Oto-Rhino-Laryngologie in Klinik und Praxis. Stuttgart, New York, Georg Thieme, 1994, pp 32–81.

3 Wright A, Davis A, Bredberg G, Ulehova L, Spencer H, Bock GR, Felix H, Iurato S, Johnsson L-G, Pauler M: Hair cell distribution in the normal human cochlea. Acta Otolaryngol Suppl (Stockh) 1987;444:4–48.

4 Retzius G: Das Gehörorgan der Wirbeltiere. II. Das Gehörorgan der Reptilien, der Vögel und der Säugetiere. Stockholm, Samson and Wallin, 1884.

5 Held H: Die Cochlea der Säuger und der Vögel, ihre Entwicklung und ihr Bau; in Bethe A, von Bergmann G, Embden G, Ellinger A (eds): Handbuch der normalen und pathologischen Physiologie. Berlin, J. Springer, 1926, vol 11: Receptionsorgane I, pp 467–534.

6 Guild SR: The width of the basiliar membrane. Science 1927;65:67–69.

7 Iurato S: Precisazioni sulla struttura microscopica e sulle variazioni diametriche della membrana basilare dell'organo del Corti. Anat Anz 1958;105:275–281.

8 Bredberg G: Cellular pattern and nerve supply of the human organ of Corti. Acta Otolaryngol Suppl (Stockh) 1968;236:6–135.

9 Coleman JW: Hair cell loss as a function of age in the normal cochlea of the guinea pig. Acta Otolaryngol (Stockh) 1976;82:33–40.

10 Keithley EM, Feldman ML: Spiral ganglion cell counts in an age-graded series of rat cochleas. J Comp Neurol 1979;188:429–442.

11 Iurato S: Submicroscopic structure of the membranous labyrinth. II. The epithelium of Corti's organ. Z Zellforschung 1961;53:259–298.

12 Engström H, Ades HW, Hawkins JE: Structure and functions of the sensory hairs of the inner ear. J Acoust Soc Am 1962;34:1356–1363.

13 Spoendlin H: Innervation densities of the cochlea. Acta Otolaryngol (Stockh) 1972;73:235–248.

14 Morrison D, Schindler RA, Wersall J: A quantitative analysis of the afferent innervation of the organ of Corti in guinea pig. Acta Otolaryngol (Stockh) 1975;79:11–23.

15 Nomura Y: Nerve fibers in the human organ of Corti. Acta Otolaryngol (Stockh) 1976;82:317–324.

16 Spoendlin H: Innervation patterns in the organ of Corti of the cat. Acta Otolaryngol (Stockh) 1969;67:239–254.

17 Liberman MC: Morphological differences among radial afferent fibers in the cat cochlea: An electron-microscopic study of serial sections. Hear Res 1980;3:45–63.

18 Nadol JB Jr: Comparative anatomy of the cochlea and auditory nerve in mammals. Hear Res 1988;34:253–266.

19 Felix H, Gleeson MJ, Johnsson LG, Pollak A: The cochlear neurons in humans; in Iurato S, Veldman JE (eds): Progress in Human Auditory and Vestibular Histopathology. Amsterdam, Kugler Publications, 1997, pp 73–79.
20 Nadol JB Jr: Serial section reconstruction of neural poles of hair cells in the human organ of Corti. I. Inner hair cells. Laryngoscope 1983;93:599–614.
21 Smith CA, Sjöstrand FS: A synaptic structure in the hair cells of the guinea pig cochlea. J Ultrastruct Res 1961;5:184–192.
22 Kimura RS: The ultrastructure of the organ of Corti. Int Rev Cytol 1975;42:173–222.
23 Fernandez C: The innervation of the cochlea (guinea pig). Laryngoscope 1951;61:1152–1172.
24 Dunn RA, Morest DK: Receptor synapses without synaptic ribbons in the cochlea of the cat. Proc Natl Acad Sci USA 1975;72:3599–3603.
25 Smith CA: Innervation pattern of the cochlea. The internal hair cell. Ann Otol Rhinol Laryngol 1961;70:1–24.
26 Spoendlin H: The organization of the cochlear receptor. Fortschr Hals Nasen Ohrenheilkd 1966; 13:1–227.
27 Kimura RS, Schuknecht HF, Sando I: Fine morphology of the sensory cells in the organ of Corti of man. Acta Otolaryngol (Stockh) 1964;58:390–408.
28 Kimura RS: Hairs of the cochlear sensory cells and their attachment to the tectorial membrane. Acta Otolaryngol 1966;61:55–72.
29 Takasaka T, Shinkawa H: Serial section reconstruction of the guinea pig outer hair cells as studied with a high-voltage electron microscope and a computer-graphic display. Acta Otolaryngol Suppl (Stockh) 1987;435:7–20.
30 Engström H, Ades HW: The ultrastructure of the cochlea; in Friedmann I (ed): Ultrastructure of Animal Tissues and Organs. Amsterdam, North Holland Publishing Company, 1973, pp 83–152.
31 Nadol JB Jr: Serial section reconstruction of the neural poles of hair cells in the human organ of Corti. II. Outer hair cells. Laryngoscope 1983;93:780–791.
32 Nadol JB Jr: Incidence of reciprocal synapses on outer hair cells of the human organ of Corti. Ann Otol Rhinol Laryngol 1984;93:247–250.
33 Pujol R, Carlier E: Cochlear synaptogenesis after sectioning the efferent bundle. Brain Res 1982; 255:151–154.
34 Smith CA: Innervation of the cochlea of the guinea pig by use of the Golgi stain. Ann Otol Rhinol Laryngol 1975;84:443–458.
35 Simmons DD: Light- and Electron-Microscopic Reconstructions of Labeled Afferent Fibers in the Adult Cat Cochlea; thesis Harvard University, Cambridge, 1986.
36 Lenoir M, Shnerson A, Pujol R: Cochlear receptor development in the rat with emphasis on synaptogenesis. Anat Embryol (Berl) 1980;160:253–262.
37 Bodian D: Synapses involving auditory nerve fibers in primate cochlea. Proc Natl Acad Sci USA 1978;75:4582–4586.
38 Howe HA: The reaction of the cochlear nerve to destruction of its end organs: A study on deaf albino cats. J Comp Neurol 1935;62:73–79.
39 Guild SR: Correlations of histologic observations and the acuity of hearing. Acta Otolaryngol (Stockh) 1932;17:207–249.
40 Hinojosa R, Seligsohn R, Lerner SA: Ganglion cell counts in the cochleae of patients with normal audiograms. Acta Otolaryngol (Stockh) 1985;99:8–13.
41 Gacek RR, Rasmussen GL: Fiber analysis of the statoacoustic nerve of guinea pig, cat, and monkey. Anat Rec 1961;139:455–463.
42 Rasmussen AT: Studies of the VIIIth cranial nerve of man. Laryngoscope 1940;50:67–83.
43 Spoendlin H: The afferent innervation of the cochlea; in Naunton RF, Fernandez C (eds): Evoked Electrical Activity in the Auditory Nervous System. New York, Academic Press, 1978, pp 21–41.
44 Kiang NY, Liberman MC, Gage JS, Northrop CC, Dodds LW, Oliver ME: Afferent innervation of the mammalian cochlea; in Keynes RD, Madrell SHP (eds): Comparative Physiology of Sensory Systems. Cambridge University Press, 1984; Cambridge, UK, pp 143–161.
45 Arnold W, Wang JB: The spiral ganglion of the rhesus monkey. Laryngol Rhinol Otol (Stuttg) 1983;62:371–377.

46 Ota CY, Kimura RS: Ultrastructural study of the human spiral ganglion. Acta Otolaryngol (Stockh) 1980;89:53–62.
47 Arnold W: Myelination of the human spiral ganglion. Acta Otolaryngol Suppl (Stockh) 1987; 436:76–84.
48 Kellerhals B: The morphology of the ganglion spirale cochleae. Acta Otolaryngol Suppl (Stockh) 1967;226:1–78.
49 Ross MD, Burkel W: Multipolar neurons in the spiral ganglion of the rat. Acta Otolaryngol (Stockh) 1973;76:381–394.
50 Rosbe KW, Burgess BJ, Glynn RJ, Nadol JB Jr: Morphologic evidence for three cell types in the human spiral ganglion. Hear Res 1996;93:120–127.
51 Tylstedt S, Kinnefors A, Rask-Andersen H: Neural interaction in the human spiral ganglion: A TEM study. Acta Otolaryngol (Stockh) 1997;117:505–512.
52 Kimura RS, Ota CY, Takahashi T: Nerve fiber synapses on spiral ganglion cells in the human cochlea. Ann Otol Rhinol Laryngol 1979;88 (suppl 62):1–17.
53 Rask Andersen H, Tylstedt S, Kinnefors A, Schrott Fischer A: Nerve fibre interaction with large ganglion cells in the human spiral ganglion: A TEM study. Auris Nasus Larynx 1997;24:1–11.
54 Rattay F, Lutter P, Felix H: A model of the electrically excited human cochlear neuron. I. Contribution of neural substructures to the generation and propagation of spikes. Hear Res 2001;153:43–63.

Dr. Heidi Felix, ENT Department, University Hospital of Zürich,
Frauenklinikstrasse 24, CH–8091 Zürich (Switzerland)
Tel. +41 1 255 58 05, Fax +41 1 255 89 59, E-Mail felix@orl.usz.ch

Felix D, Oestreicher E (eds): Rational Pharmacotherapy of the Inner Ear.
Adv Otorhinolaryngol. Basel, Karger, 2002, vol 59, pp 11–17

Neurotransmission in the Human Labyrinth

Anneliese Schrott-Fischer[a], *Keren Kammen-Jolly*[a], *Arne W. Scholtz*[a], *Wei-jia Kong*[b], *Michel Eybalin*[c]

[a] Department of Otolaryngology, University of Innsbruck, Innsbruck, Austria;
[b] Department of Otolaryngology, Hua-Zhong University of Science and Technology, Hua-Zhong, China;
[c] Inserm U 254 Neurobiologie de l'Audition, Montpellier, France

Abstract

Different neuroactive substances have been found in the efferent pathways of both the olivocochlear and vestibular systems. In the present study, the distribution and role of three neurotransmitters, choline acetyltransferase (ChAT), gamma aminobutyric acid (GABA), and enkephalin were investigated in the human labyrinth of 4 normal-hearing individuals. Immunohistochemical studies in human inner ear research, however, face a problem of procuring well-preserved specimens with maintained neurotransmitter antigenicity and morphology. Methods and findings are reported and discussed.

The membranous inner ear and its associated sensory end-organs, including the cochlea, the sacculus, the utriculus, and the semicircular canals, are encased within the temporal bone. While the cochlea may transduce and transmit sensory signals related to audition, the vestibular end-organs are involved in sensing movement and position [1]. The presence of efferent components in the vestibulocochlear system was first demonstrated by Rasmussen and Gacek [2]. Both cochlear and vestibular end-organs receive efferent innervation that originates from the brainstem. The cochlear efferents consist of two separate systems called the lateral olivocochlear (LOC) and the medial olivocochlear (MOC) [3]. The LOC innervates the inner spiral bundles beneath inner hair cells whereas the MOC synapses directly with outer hair cells. Type I auditory neurons form

90–95% of the spiral ganglion and contact inner hair cells exclusively. Type II neurons account for the 5–10 remaining percent of spiral ganglion neurons and innervate outer hair cells [4]. The Vestibular efferents originate from three groups of neurons in the brainstem and either establish an axo-dendritic synapse with afferent calyces surrounding type I hair cells or a direct synapse with type II vestibular hair cells [5–8].

It is currently accepted that the synaptic transmission in the efferent vestibular system of mammals is chemical in nature [9]. Findings have shown that more than a single neuroactive substance can coexist in the same nerve terminal and that one neurotransmitter may act at different postsynaptic receptors. Different neuroactive substances have been found in the efferent pathways of both the olivocochlear and vestibular systems. The first identified members of the large group of endogenous opioid peptides called endorphins are leucine(leu)- and methionine(met)-enkephalin. They originate in varying proportions from all three known opioid peptide precursors [10]. Their distribution and function have not yet been clarified in the vestibulocochlear system.

Acetylcholinesterase (AchE) is considered a major neurotransmitter of the MOC and LOC system in the mammalian cochlea [11]. Choline acetyltransferase (ChAT) is the synthesizing enzyme for acetylcholine and has been shown in guinea pig [12] and rat cochlea [13, 14]. In agreement with these findings, the vestibular efferent neurons of the squirrel monkey have also been shown to contain ChAT [15]. Immunohistochemical data also suggest the possible neurotransmission of GABA in the efferent innervation of the cochlea. GABA has been described in varicosities and fibers within the inner spiral bundle of various mammalian species [16, 17]. However, in a study conducted on the primate squirrel monkey, GABA-like-immunoreactivity was found to be distributed throughout the entire length of the organ of Corti as well as the vestibular endorgans [18]. The localization patterns of GABA in the vertebrate vestibular periphery vary according to species although GABA, an afferent neurotransmitter at the hair cell level, has been implicated. The aim of this investigation was to further elucidate the distribution and meaning of three neurotransmitters, ChAT, GABA, and enkephalin, in the human labyrinth.

Material and Methods

Four temporal bones obtained from normal hearing individuals, were processed over the past five years in our laboratory using different fixation techniques. The temporal bones were obtained from the Department of Otolaryngology, the Department of Pathology. After perilymphatic perfusion with either fixative A – 4% paraformaldehyde with 0.1% glutaraldehyde in 0.1 *M* phosphate-buffered saline (PBS), pH 7.4 – or fixative C – 5% glutaraldehyde in 0.1 *M* cacodylate buffer with 1% sodium metabisulfate, pH 7.2 – the oval and round windows were sealed with ear wax for the GABA study. The bone around the membranous labyrinth was removed by drilling in cold PBS (4 °C), and the five end-organs of the vestibule and the organ of Corti were dissected under a stereomicroscope. The specimens were then processed for immunocytochemistry using antibodies against ChAT or GABA.

Results

ChAT immunoreactivity was found in the organ of Corti along inner spiral, inner tunnel spiral, and outer spiral fibers. Just below the outer hair cells, immunostaining was distributed along efferent nerve endings. In contrast, adjacent afferent nerve fibers appeared consistently free of peroxidase reaction product. Staining could not be detected in any of our negative controls using normal rabbit serum. In the human vestibule, the distribution of ChAT was found within all five end-organs, including both the cristae and the maculae. Type I and type II sensory hair cells, afferent nerve calyces, and supporting cells did not appear immunoreactive. In the vestibular epithelia, reactivity was confined to regions of neurosensory epithelium. The majority of ChAT immunostained components were located within the lower one-third portion of neurosensory epithelia along basal regions of sensory hair cells. Results gathered on rat temporal bones were in agreement with our findings of ChAT reactivity in humans. In all cochleae examined, immunostaining for GABA was detected along efferent endings at the base of both inner and outer hair cells. Reactivity could be observed at the vestibular periphery of the human inner ear but was confined to the vesiculated nerve endings and unmyelinated nerve fibers. Type I and type II hair cells, afferent calyces, and myelinated nerve fibers did not demonstrate reactivity. GABA-like immunoreactivity in the rat was comparable to patterns of distribution found in humans. Colocalized patterns of met- and leu-enkephalin-like immunoreactivity were specific to the organ of Corti and were not evidenced in end-organs of the vestibule, either the cristae or the maculae. At the electron-microscopic level, positive reactivity was verified along the intraganglionic spiral bundle, the inner spiral bundle, the tunnel spiral bundle, and within granulated nerve endings (compatible with efferents) beneath inner and outer hair cells. Immunostaining of adjacent afferent nerve fibers was not found (fig. 1).

Discussion

Our findings of ChAT immunoreactivity in the human cochlea agree with previous studies performed in the rat [13, 16, 19]. Furthermore, histochemical results by various authors demonstrate acetylcholinesterase in perikarya forming

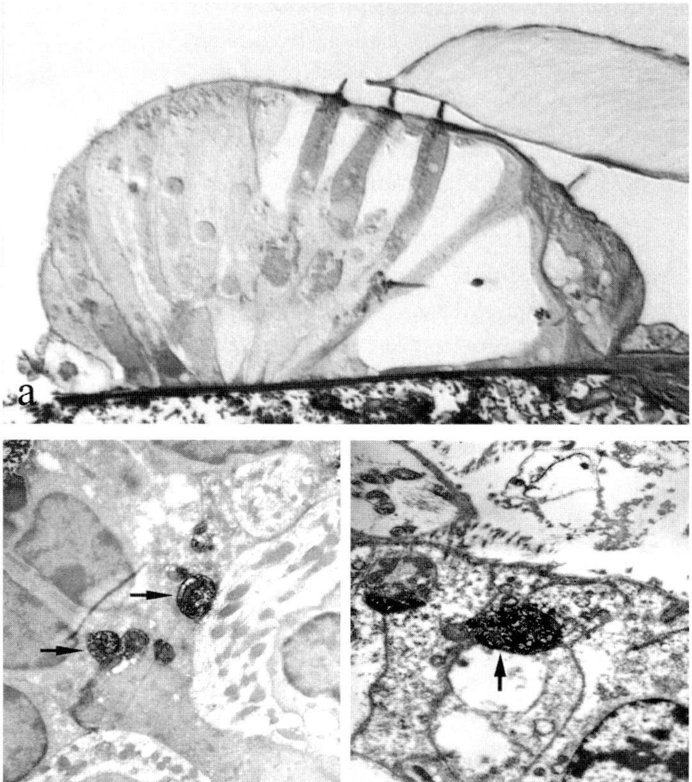
a
b
c
d

both the LOCs and MOCs in the superior olivary complex. The data accumulated thus suggest that acetylcholine in the human organ of Corti is an efferent neurotransmitter operative in both the LOC and MOC system. ChAT distribution in humans also coincides with that in other investigated mammals [9]. Likewise, ChAT immunostaining found in the human vestibular periphery provides evidence that Ach may be a candidate neurotransmitter in the efferent vestibular system. Hence, it is demonstrated that in human, cholinergic efferents innervate both type I and type II vestibular hair cells making postsynaptic and presynaptic contacts respectively. GABA has been demonstrated in the efferent system of both the human cochlea and the human vestibule. Although there are discrepancies between GABA reactivity in the chicken and guinea pig, findings in the rat and humans agree. Nonetheless, vesiculated efferents that are positive for GABA immunostaining are shown to innervate exclusively at afferent calyces and thus differ from ChAT immunoreactive efferents. Previous studies have demonstrated enkephalin-like-immunoreactivity in the lateral olivocochlear neurons of the cat and guinea pig [12, 20–22]. It seems likely that enkephalins participate in the modulation of auditory processes [23]. The absence of enkephalins in the vestibular end-organs may add further support to electrophysiological studies that suggest enkephalins may inhibit transmission in the medial vestibular nucleus type I neurons thereby controlling vestibulo-ocular reflexes [24, 25]. Immunohistochemical studies in human inner ear research are faced with the problem of procuring well-preserved specimens with maintained antigenicity. The time interval between death and fixation of the inner ear tissues is decisive for preserving both the cytoarchitecture and the antigenicity of the sensory end-organs and nerve components. Typically, to prepare human inner ear tissue, decalcification was time-consuming and some neurotransmitter antigens became undetectable. Anoxia can lead to degeneration of cells and neural elements that may generate false-negative results or false-positive (artefacts) interpretations. Temporal bones fixed more than 5 h postmortem then were inadequate due to autolysis during the long immersion period in the decalcifying solution. In addition, different kinds of fixatives were needed for antigen preservation of different neurotransmitters. Insufficient tissue

Fig. 1. ***a*** ChAT-like immunoreactivity in the human organ of Corti. ChAT-like positive staining was found beneath the inner hair cells in tunnel crossing fibers and efferent nerve endings at the base of outer hair cells. × 320. ***b*** Immunoelectron micrograph of the utricular macula with ChAT like-immunoreactivity in the efferent fibers (arrows). Scale bar: 1 μm. ***c*** Electron micrograph at the basal area of the neurosensory epithelium of a human utricular macula with GABA-like immunoreactivity. The nonvesiculated afferent fibers are void of reactivity. Scale bar: 1 μm. ***d*** Electron micrograph of the sensory epithelium of the human saccule where met- and leu-enkephalin like-immunoreactivities were absent. Scale bar: 3 μm.

fixation led to leaching of inadequately fixed antigens and did not offer good preservation of morphology. Overfixation could result in masking or denaturation of the antigen to be detected, which is problematic as well. By applying preembedding immunostaining techniques, we were able to circumvent the decalcification procedure altogether thus enabling rapid evaluation of the tissue. Essentially then, using perilymphatic perfusion and immersion in combination with the microdissection technique, we have achieved high quality preservation of the morphology of both neurotransmitter antigenicity and morphology of the human inner ear.

References

1 Beitz AJ, Anderson JH: Neurochemistry of the vestibular system. Boca Raton, CRC Press, 2000.

2 Rasmussen GL, Gacek RR: Concerning the question of the efferent fiber component of the vestibular nerve of the cat. Anat Rec 1958;130:361–362.

3 Warr WB: Olivocochlear and vestibular efferent neurons of the feline brain stem: Their location, morphology and number determined by retrograde axonal transport and acetylcholinesterase histochemistry. J Comp Neurol 1975;161:159–181.

4 Spoendlin H: Innervation densities of the cochlea. Acta Otolaryngol 1972;73:235–248.

5 Wersall J: Studies on the structure and innervation of the sensory epithelium of the cristae ampullaris in the guinea pig. A light and electron microscopic investigation. Acta Otolaryngol 1956;(Suppl 126):1–85.

6 Smith C, Rasmussen PE: Nerve endings in the maculae and cristae of the chinchilla vestibule, with special reference to the efferents. Third Symposium on the Role of the Vestibular Organs in Space Exploration. NASA SP-152. Washington, NASA SP-152, 1967, pp 183–201.

7 Iurato S, Luciano L, Pannese E, et al: Efferent vestibular fibers in mammals: Morphological and histochemical aspects. Prog Brain Res 1972;37:429–443.

8 Wersall J, Bagger-Sjoback D: Morphology of the vestibular sense organ; in Kornhuber HH (ed): Handbook of Sensory Physiology. Vestibular System. Basic Mechanism. New York, Springer, 1974, vol 9, pp 123–170.

9 Kong WJ, Hussl B, Thumfart WF, et al: Ultrastructural localization of ChAT-like immunoreactivity in the human vestibular periphery. Hear Res 1998;119:96–103.

10 Scholtz AW, Kanonier G, Schrott-Fischer A: Immunohistochemical investigation of enkephalins in the human inner ear. Hear Res 1998;118:123–128.

11 Eybalin M: Neurotransmitters and neuromodulators of the mammalian cochlea. Physiol Rev 1993;73:309–373.

12 Altschuler RA, Hoffman DW, Reeks KA, et al: Localization of dynorphin B-like and alpha-neoendorphin-like immunoreactivities in the guinea pig organ of Corti. Hear Res 1985;17:249–258.

13 Eybalin M, Pujol R: Immunolocalization of choline acetyltransferase in 2 types of efferent synapses of the organ of Corti. C R Acad Sci III 1985;301:679–684.

14 Eybalin M, Pujol R: Choline acetyltransferase (ChAT) immunoelectron microscopy distinguishes at least three types of efferent synapses in the organ of Corti. Exp Brain Res 1987;65:261–270.

15 Carpenter MB, Chang L, Pereira AB, et al: Comparisons of the immunocytochemical localization of choline acetyltransferase in the vestibular nuclei of the monkey and rat. Brain Res 1987; 418:403–408.

16 Eybalin M, Altschuler RA: Immunoelectron microscopic localization of neurotransmitters in the cochlea. J Electron Microsc Tech 1990;15:209–224.

17 Eybalin M, Parnaud C, Geffard M, et al: Immunoelectron microscopy identifies several types of GABA-containing efferent synapses in the guinea-pig organ of Corti. Neuroscience 1988;24:29–38.

18 Thompson GC, Cortez AM, Igarashi M: GABA-like immunoreactivity in the squirrel monkey organ of Corti. Brain Res 1986;372:72–79.
19 Dannhof BJ, Roth B, Bruns V: Anatomical mapping of choline acetyltransferase (ChAT)-like and glutamate decarboxylase (GAD)-like immunoreactivity in outer hair cell efferents in adult rats. Cell Tissue Res 1991;266:89–95.
20 Fex J, Altschuler RA: Enkephalin-like immunoreactivity of olivocochlear nerve fibers in cochlea of guinea pig and cat. Proc Natl Acad Sci USA 1981;78:1255–1259.
21 Altschuler RA, Fex J, Parakkal MH, et al: Colocalization of enkephalin-like and choline acetyltransferase-like immunoreactivities in olivocochlear neurons of the guinea pig. J Histochem Cytochem 1984;32:839–843.
22 Eybalin M, Pujol R: Immunofluorescence with Met-enkephalin and Leu-enkephalin antibodies in the guinea pig cochlea. Hear Res 1984;13:135–140.
23 Code RA, Carr CE: Enkephalin-like immunoreactivity in the chick brainstem: Possible relation to the cochlear efferent system. Hear Res 1995;87:69–83.
24 Kawabata A, Sasa M, Ujihara H, et al: Inhibition by enkephalin of medial vestibular nucleus neurons responding to horizontal pendular rotation. Life Sci 1990;47:1355–1363.
25 Highstein SM: The central nervous system efferent control of the organs of balance and equilibrium. Neurosci Res 1991;12:13–30.

Dr. Anneliese Schrott-Fischer, Department of Otolaryngology,
University of Innsbruck, Anichstrasse 35, A-A-6020 Innsbruck (Austria)
Tel. +43 512 504 3174, Fax +43 512 504 5218, E-Mail Annelies.Schrott@uibk.ac.at

Felix D, Oestreicher E (eds): Rational Pharmacotherapy of the Inner Ear.
Adv Otorhinolaryngol. Basel, Karger, 2002, vol 59, pp 18–25

Different Action of Memantine and Caroverine on Glutamatergic Transmission in the Mammalian Cochlea

Elmar Oestreicher[a,b], *Klaus Ehrenberger*[c], *Dominik Felix*[b]

[a] ENT Department, Technical University of Munich, Germany;
[b] Division of Neurobiology, University of Berne, Switzerland;
[c] ENT Department, University of Vienna, Austria

Abstract

Glutamate is the major transmitter candidate between inner hair cells and the afferent neurons of the mammalian cochlea. We investigated the action of memantine (1-amino-3,5-dimethyl-adamantane) and the quinoxaline derivative caroverine [1-diethylaminoethyl-3,8-(*p*-methoxybenzyl)-1,2-dihydro-quinoxaline-dione] on the glutamatergic transmission in the guinea pig cochlea utilizing extracellular recording techniques and microiontophoretic ejection of substances. While memantine was able to inhibit the NMDA (N-methyl-*D*-aspartate)-induced firing, the AMPA (α-amino-3-hydroxy-5-methylisoxazole-4-propionic acid)-stimulated activity was unaffected. In contrast, caroverine could block both NMDA- as well as AMPA-induced firing. As memantine and caroverine are currently in clinical use, these substances could be introduced to the treatment of several cochlear disorders.

Neurotransmission in the mammalian cochlea between inner hair cells and the primary auditory neurons (type I spiral ganglion neurons) is most probably mediated by glutamate [1]. Both classes of ionotropic glutamate receptor subtypes, N-methyl-*D*-aspartate (NMDA) and α-amino-3-hydroxy-5-methylisoxazole-4-propionic acid (AMPA) [review in 2], seem to be present on the afferent nerve fibers of inner hair cells [1, 3]. Besides its role as a fast excitatory neurotransmitter, glutamate has neurotoxic properties in the central nervous system. This occurs when it is released in excess or when its recycling mechanism breaks down. Glutamate-mediated neurotoxicity has been implicated in pathophysiological conditions such as ischemia, hypoglycemia, anoxia, trauma, neurodegenerative disorders and aging [4]. In the cochlea, excitotoxicity has been described to play a role in the pathology of ischemia and noise trauma [5]. Glutamate antagonists like the AMPA antagonist 6,7-dinitroquinoxaline-2,3-dione and – to a lesser degree – the NMDA antagonist *D*-2-amino-5-phosphonopentanoate have been shown to protect afferent cochlear neurons from excitotoxicity [6].

Since both classes of ionotropic glutamate receptors may be involved in several inner ear diseases like presbyacusis, progressive hearing loss or tinnitus [5, 7], antiglutamatergic drugs could be useful as a therapeutic strategy in different cochlear disorders.

In an earlier report we described the glutamate antagonism of caroverine, a quinoxaline derivative, on the neurotransmission of inner hair cells in guinea pigs [8]. The efficacy of caroverine for the treatment of patients suffering from tinnitus has been demonstrated recently in a double-blind study [9].

The long-known antiparkinsonian and antispastic substance memantine has been characterized as a noncompetitive NMDA receptor blocker [10, 11]. Memantine protects cortical neuron cultures from the toxicity of glutamate and NMDA [12, 13] and has antihypoxic properties in vitro and in vivo [14]. Memantine is currently used in the treatment of Parkinson's disease [15] and psychiatric disorders [16]. Recently, the NMDA antagonism of memantine could be demonstrated in the cochlea [17].

The aim of this study was to characterize the receptor selectivity of both caroverine and memantine on the glutamatergic neurotransmission of cochlear inner hair cells.

Methods

Adult female pigmented guinea pigs (600–870 g) were anesthetized with a combination of xylazine (0.4 ml/kg; Rompun, Bayer, Leverkusen, Germany) and droperidol/fentanyl (1.3 ml/kg; Innovar-Vet, Pitman-Moore, Mundelein, Ill., USA). To assure a constant deep level of anesthesia, additional low doses (0.1 ml) were administered periodically. Body temperature was monitored and maintained within physiological limits. After tracheotomy to allow artificial respiration during the experiment, the auditory bulla was approached ventrolaterally and opened. With the aid of a stereoscopic microscope a small opening was drilled in the cochlear bone right over the pigmented stria vascularis and the ligamentum spirale of the third or fourth turn. For recordings and administration of substances a 5-barrel glass microelectrode with a tip diameter between 2 and 3 μm was inserted through the basal border of the stria vascularis and then driven almost parallel to the tectorial membrane. The subsynaptic region was reached at a depth of about 200–280 μm as indicated by typical phasic activity. Recording of the extracellular action potentials was performed with standard electrophysiological equipment using a 2 *M* NaCl-filled barrel, displayed online via an oscilloscope and printed out with a thermoprinter. For later analysis, data were additionally recorded on an audiotape.

Substances applied with appropriate anionic or cationic currents by the other four channels of the microelectrode included: *L*-glutamic acid (Sigma, 0.1 *M*, pH 7.5, adjusted with NaOH), NMDA (Sigma, 0.1 *M*, pH 8, NaOH), AMPA (Tocris, 0.01 *M*, pH 8, NaOH), memantine-HCl (RBI, 0.005 *M*, pH 3.8, HCl) and caroverine (Donauchemie Linz, 0.05 *M*, natural pH 6.0).

The care and use of the animals reported on in this study were approved by the State Animal Care and Use Committee, Department Service of Veterinarians, Berne, Switzerland, provided by grant No. 33/98.

Results

The iontophoretic application of glutamate or glutamate agonists like NMDA or AMPA enhanced the spontaneous firing of afferent fibers of inner hair cells. This was a consistent finding on all 34 fibers tested.

In a first set of experiments, we tested the ability of memantine and caroverine on the spontaneous and the glutamate-induced firing of afferent dendrites of inner hair cells. Memantine and to a lesser degree caroverine exhibited a suppressing effect on spontaneous firing. Both, memantine and caroverine, antagonized the stimulating effect of glutamate in a reversible manner on all fibers tested (n = 9). Simultaneous application of both substances led to an enhanced reduction of the glutamate-induced firing.

In a second set of experiments we were interested in the glutamate receptor subtypes responsible for these effects. Therefore, the effects of memantine and caroverine were tested on NMDA- or AMPA-stimulated fibers (n = 25). A reversible depressant effect of memantine and caroverine could be observed on NMDA-stimulated fibers. Figure 1a illustrates that an equal amount of substances, ejected with the same amount of current (50 nA), showed a stronger depressant action of memantine than of caroverine. In contrast, the AMPA-induced firing of afferent fibers (fig. 1b) could not be blocked by memantine. The enhanced firing, however, was affected by caroverine.

Figure 2 illustrates a different approach to the possible interaction between memantine and AMPA- or NMDA-induced firing. Whereas in figure 1 NMDA or AMPA were ejected continuously over a longer period, the glutamate receptor subtype agonists were administered over shorter periods of 10–30 s with an equal amount of current (15 nA). The selective antagonism of memantine on NMDA- but not on AMPA-induced activity is again clearly demonstrated. After cessation of the ejection of memantine the excitatory action of NMDA is recovered within minutes.

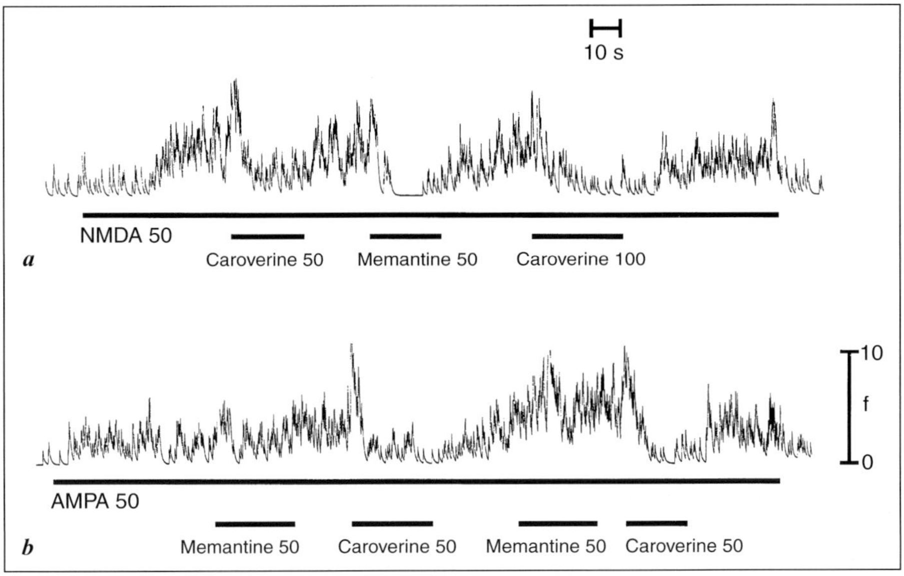

Fig. 1. Recording of a single afferent fiber, which is first activated by 50 nA of NMDA (***a***) and secondly with AMPA (***b***). Under NMDA stimulation, caroverine shows only weak antagonism, while memantine blocks the NMDA-induced firing effectively. In contrast, the AMPA-induced activity could be blocked by caroverine, whereas memantine exhibited no effect on the AMPA-induced activity. In the present and the following figure, fiber activity is demonstrated on integrated time frequency curves. Administration of the substances is indicated by bars beneath the frequency curves (f = frequency in hertz (H)).

Discussion

Both substances, memantine and caroverine, were able to block glutamate-induced firing of afferent fibers of inner hair cells in guinea pigs. They differ, however, in interacting with the glutamate receptor subtypes NMDA and AMPA. As expected from earlier reports on recombinant NMDA receptors, binding and patch-clamp experiments [10, 11, 18–20], memantine showed a selective blockade of NMDA-stimulated activity, while the AMPA-induced firing of afferent fibers of cochlear inner hair cells was unaffected. Thus, in the mammalian cochlea memantine can be characterized as an NMDA receptor blocker, as reported earlier [17].

The antiglutamatergic properties of caroverine, which was first considered to be an unspecific calcium channel blocker [21], for the mammalian cochlea have already been reported [8]. The results of this study demonstrate that caroverine is a potent inhibitor of AMPA- and to a lesser degree of NMDA-induced activity of

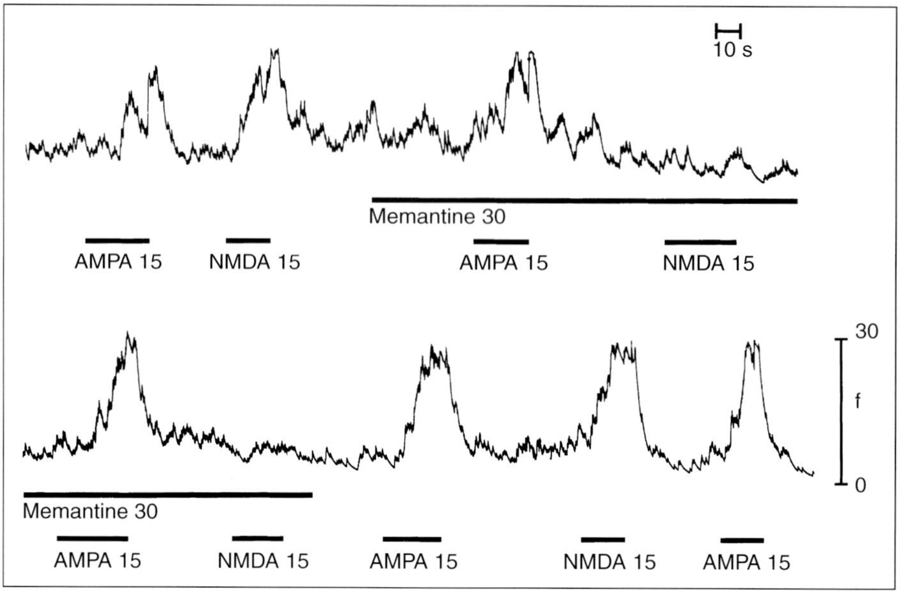

Fig. 2. Consecutive recording of an afferent nerve fiber with short activation intervals by NMDA (15 nA) and AMPA (15 nA) before, during and after application of memantine (30 nA). Note that with memantine only AMPA-induced firing could be observed, whereas the action of NMDA was totally blocked.

afferent cochlear neurons. This seems to be similar to other quinoxaline derivatives: 6-cyano-7-nitroquinoxaline-2,3-dione and 6,7-dinitroquinoxaline-2,3-dione have non-NMDA and in higher concentrations also NMDA-antagonistic potentials [22, 23].

The enhanced antiglutamatergic effect during simultaneous application of both substances as observed in this study could be explained by the lesser degree of NMDA blockade of caroverine. Thus, remaining unblocked NMDA receptors could be blocked by memantine and would result in a stronger antiglutamatergic effect.

This synergistic effect opens the possibility to combine both substances to increase the efficacy of blocking glutamatergic neurotransmission for therapeutic reasons.

In the mammalian cochlea excitotoxicity affects the afferent neurons postsynaptically to the inner hair cells. This may occur in particular in noise trauma and presbyacusis [5].

The results of this study make it reasonable to consider both substances as antiglutamatergic drugs for such inner ear diseases like noise trauma, sudden hearing loss or tinnitus. As they are already currently in clinical use, they could be introduced to this new therapeutic field very easily.

The safety and efficacy of caroverine was documented in several clinical studies for the treatment of alcohol withdrawal [24], cerebral hypoxia [25] and tinnitus [9].

In contrast to other NMDA channel blockers like phencyclidine or ketamine, which have severe neuropsychiatric side effects [4], memantine is very well tolerated and in clinical use for the treatment of Parkinson's disease [15], dementia [26] and psychiatric disorders [16]. The memantine concentration found in the cerebrospinal fluid during therapeutic use suggests that its primary site of action is the NMDA receptor [27].

Why memantine appears to induce fewer and less profound effects on perception or consciousness [26] than other NMDA channel blockers like phencyclidine or ketamine could have different reasons. It has been supposed that some NMDA channel blockers like memantine may be safer because they block high levels of glutamate, while sparing synaptic responses [28]. This is supported by the lower unblocking rate constants of phencyclidine and ketamine compared to that of memantine [19, 28].

The partial trapping channel block of memantine, leaving a portion of channels unblocked during synaptic activity, has recently been published [19] and could also account for the safety of memantine.

Taken together, memantine and caroverine are well tolerated and efficient antiglutamatergic drugs already in clinical use. They block preferably different ionotropic glutamate receptors. The different selectivity of caroverine and memantine to NMDA and non-NMDA receptors opens the possibility to block effectively both groups of ionotropic glutamate receptors. In the mammalian cochlea, memantine and caroverine are effective in blocking the glutamatergic transmission and thus could be of great value as new therapeutic drugs for inner ear diseases.

Together with dopamine agonists, which protect afferent dendrites of inner hair cells from neurotoxicity and depress the glutamatergic neurotransmission as well [29, 30], memantine and caroverine could be part of a new generation of agents for inner ear therapy.

References

1 Eybalin M: Neurotransmitter and neuromodulators in the mammalian cochlea. Physiol Rev 1993;73:309–373.
2 Watkins JC, Olverman HJ: Agonists and antagonists for excitatory amino acid receptors. Trends Neurosci 1987;10:265–272.
3 Ehrenberger K, Felix D: Glutamate receptors in the afferent cochlear neurotransmission in guinea pigs. Hear Res 1991;52:73–80.

4 Lipton SA, Rosenberg PA: Excitatory amino acids as a final common pathway for neurologic disorders. N Engl J Med 1994;330:613–622.
5 Pujol R, Puel JL, d'Aldin CG, Eybalin M: Pathophysiology of the glutamatergic synapses in the cochlea. Acta Otolaryngol (Stockh) 1993;113:330–334.
6 Puel JL, Pujol R, Tribillac F, Ladrech S, Eybalin M: Excitatory amino acid antagonists protect cochlear auditory neurons from excitotoxicity. J Comp Neurol 1994;341:241–256.
7 Ehrenberger K, Felix D: Receptor pharmacological models for inner ear therapies with emphasis on glutamate receptors: A survey. Acta Otolaryngol (Stockh) 1995;115:236–240.
8 Ehrenberger K, Felix D: Caroverine depresses the activity of cochlear glutamate receptors in guinea pigs: In vivo model for drug-induced neuroprotection? Neuropharmacology 1992;31: 1259–1263.
9 Denk DM, Heinzl H, Franz P, Ehrenberger K: Caroverine in tinnitus treatment. A placebo-controlled blind study. Acta Otolaryngol (Stockh) 1997;117:825–830.
10 Bormann J: Memantine is a potent blocker of N-methyl-*D*-aspartate (NMDA) receptor channels. Eur J Pharmacol 1989;166:591–592.
11 Parsons CG, Gruner R, Rozental J, Millar J, Lodge D: Patch clamp studies on the kinetics and selectivity of N-methyl-*D*-aspartate receptor antagonism by memantine (1-amino-3,5-dimethyladamantane). Neuropharmacology 1993;32:1337–1350.
12 Erdö SL, Schäfer M: Memantine is highly potent in protecting cortical cultures against excitotoxic cell death evoked by glutamate and N-methyl-*D*-aspartate. Eur J Pharmacol 1991;198:215–217.
13 Weller M, Finiels-Marlier F, Paul SM: NMDA receptor-mediated glutamate toxicity of cultured cerebellar, cortical and mesencephalic neurons: Neuroprotective properties of amantadine and memantine. Brain Res 1993;613:143–148.
14 Seif el Nasr M, Peruche B, Rossberg C, Mennel HD, Krieglstein J: Neuroprotective effect of memantine demonstrated in vivo and in vitro. Eur J Pharmacol 1990;185:19–24.
15 Fischer PA, Jacobi P, Schneider E, Schönberger B: Die Wirkung intravenöser Gaben von Memantine bei Parkinson-Kranken. Drug Res 1977;27:1487–1589.
16 Weller M, Kornhuber J: Pathophysiology and therapy of malignant neuroleptic syndrome. Nervenarzt 1991;63:645–655.
17 Oestreicher E, Arnold W, Ehrenberger K, Felix D: Memantine suppresses the glutamatergic neurotransmission of mammalian inner hair cells. ORL 1998;60:18–21.
18 Kornhuber J, Bormann J, Retz W, Hübers M, Riederer P: Memantine displaces [^{3}H]MK-801 at therapeutic concentrations in postmortem human frontal cortex. Eur J Pharmacol 1989;166: 589–590.
19 Blanpied TA, Boeckman FA, Aizenman E, Johnson JW: Trapping channel block of NMDA-activated responses by amantadine and memantine. J Neurophysiol 1997;77:309–323.
20 Bresink I, Benke TA, Collett VJ, Seal AJ, Parsons CG, Henley JM, Collingridge GL: Effects of memantine on recombinant rat NMDA receptors expressed in HEK 293 cells. Br J Pharmacol 1996;119:195–204.
21 Kudo Y, Shibata S: Effects of caroverine and diltiazemon synaptic responses, *L*-glutamate-induced depolarization and potassium efflux in the frog spinal cord. Br J Pharmacol 1984;83:813–820.
22 Honoré T, Davies SN, Drejer J, Fletcher EJ, Jacobsen P, Lodge D, Nielsen FE: Quinoxalinediones: Potent competitive non-NMDA glutamate receptor antagonists. Science 1988;241:701–703.
23 Kessler M, Baudry M, Lynch G: Quinoxaline derivatives are high-affinity antagonists of the NMDA receptor associated glycine sites. Brain Res 1989;489:377–382.
24 Koppi S, Eberhardt G, Haller R, König P: Calcium-channel-blocking agent in the treatment of acute alcohol withdrawal – Caroverine versus meprobamate in a randomized double-blind study. Neuropsychobiology 1987;17:49–52.
25 Saletu B, Grunberger J, Anderer P, Linzmayer L, König P: On the cerebro-protective effects of caroverine, a calcium-channel blocker and antiglutamatergic drug: Double-blind, placebo-controlled, EEG mapping and psychometric studies under hypoxia. Br J Clin Pharmacol 1996; 41:89–99.
26 Ditzler K: Efficacy and tolerability of memantine in patients with dementia syndrome. Drug Res 1991;41:773–780.

27 Kornhuber J, Quack G: Cerebrospinal fluid and serum concentrations of the N-methyl-*D*-aspartate (NMDA) receptor antagonist memantine in man. Neurosci Lett 1995;195:137–139.
28 Chen HSV, Pellegrini JW, Aggarwal SK, Lei SZ, Warach S, Jensen FE, Lipton SA: Open-channel block of N-methyl-*D*-aspartate (NMDA) responses by memantine: Therapeutic advantage against NMDA receptor-mediated neurotoxicity. J Neurosci 1992;12:4427–4436.
29 d'Aldin C, Puel JL, Leducq R, Crambes O, Eybalin M, Pujol R: Effects of a dopaminergic agonist in the guinea pig cochlea. Hear Res 1995;90:202–211.
30 Oestreicher E, Arnold W, Ehrenberger K, Felix D: Dopamine regulates the glutamatergic inner hair cell activity in guinea pigs. Hear Res 1997;107:46–52.

Dr. Elmar Oestreicher, ENT Department, Technical University of Munich,
Ismaninger Strasse 22, D–81675 Munich (Germany)
Tel. +49 89 41402370, Fax +49 89 41404853, E-Mail e.oestreicher@lrz.tum.de

Felix D, Oestreicher E (eds): Rational Pharmacotherapy of the Inner Ear.
Adv Otorhinolaryngol. Basel, Karger, 2002, vol 59, pp 26–34

Role of Substance P in the Peripheral Vestibular and Auditory System

H. Felix[a], *E. Oestreicher*[b,c], *D. Felix*[c], *K. Ehrenberger*[d]

[a] ENT Department, University Hospital Zürich, Zürich, Switzerland;
[b] ENT Department, TU Munich, Munich, Germany;
[c] Division of Neurobiology, University of Berne, Berne, Switzerland;
[d] ENT Department, University of Vienna, Vienna, Austria

Abstract

The central role of substance P (SP) has attracted growing interest in the past two decades. One of the important physiological functions of SP and other tachykinins is that of a neurotransmitter in primary afferent neurons. Recent immunocytochemical, biochemical and electrophysiological investigations on various neurotransmitters support the hypothesis that SP has a similar function in the vestibular and auditory systems of all mammals including humans. The purpose of this review is to give an overview of the distribution and concomitant physiological functions of this peptide in these sensory systems.

Among many different physiological functions the most important role of substance P (SP) is that of a neurotransmitter [for a review, see ref. 1]. In the past, findings based on a great variety of approaches have added new and fundamental information to our understanding of this peptide. The determination of the tachykinin precursors by molecular cloning and sequence analysis made it possible to define the structure of the receptors and to analyze their distribution. Many other substances have become valuable pharmacological and physiological tools in studying the role of tachykinins. Capsaicin, for example, caused a fast calcium-dependent release of SP from central and peripheral terminals of primary afferents [2]. Research on the functional role of tachykinins has therefore mainly focused on the primary sensory neurons. Furthermore, the coexistence of peptides with classical neurotransmitters suggested that peptides interact with neurons and play an important and distinct role in neuronal mechanisms. There is increasing evidence that SP may affect the function in the vestibular end-organs and auditory systems of several mammalian species. The purpose of this review is to present the distribution pattern and physiological functions of SP in vestibular and auditory systems.

The Peripheral Vestibular System

The presence and distribution pattern of SP in the vestibular peripheral system is illustrated mainly by immunohistochemical findings. Ontogenetic studies have shown that SP immunoreactivity occurred relatively early during gestation. Its occurrence during different periods of gestation suggests that the peptide may play some crucial role during development. In rats, SP can be detected in the afferent nerve fibers of the utricular and saccular maculae on gestational day 19 [3]. There are no physiological data, however, on whether the afferent nerve fibers are already functioning at this stage. The fact that compact myelin of the dendrites was observed on the same day of incubation suggests that the morphological basis of functional activity is present around this period. SP present in the epithelium at critical stages of development may therefore be involved in the maturation of vestibular tachykinin receptors [3].

Nevertheless, the occurrence of SP in the afferent nerve fibers raises the question whether the peptide is involved in sensory neurotransmission. Most histochemical studies have dealt with the distribution pattern of SP-like immunoreactivity in the vestibular ganglion of different animal species as well as humans [4–6]. An important question related to this topic was the correlation between SP immunoreactivity and the different ganglion cell types. In the guinea pig more than 80% of vestibular ganglion cells showed SP-like immunoreactivity [7]. Similar immunocytochemical features of SP were reported in small cells of primate and cat vestibular ganglions [8]. In humans, SP immunoreactivity was located in 10–15% of predominantly small ganglion cells with a diameter of about 20 μm, as illustrated in figure 1A. The thin SP-labeled nerve fibers most probably represent the nerve processes (fig. 1B) of the labeled small ganglion cells [9]. The distribution of SP in the vestibular ganglion, which is confined to a selective portion of small ganglion cells, and the presence of specific mRNA encoding the precursor of the peptide suggest that SP is synthesized in those cells where it may act as a primary sensory transmitter.

An interesting observation has been made in human vestibular ganglions. In contrast to mammals, vestibular ganglion cells in humans do not have myelin sheaths and many ganglion cells have close contact with an adjacent ganglion cell [10]. Electrophysiological studies have shown that interaction with and

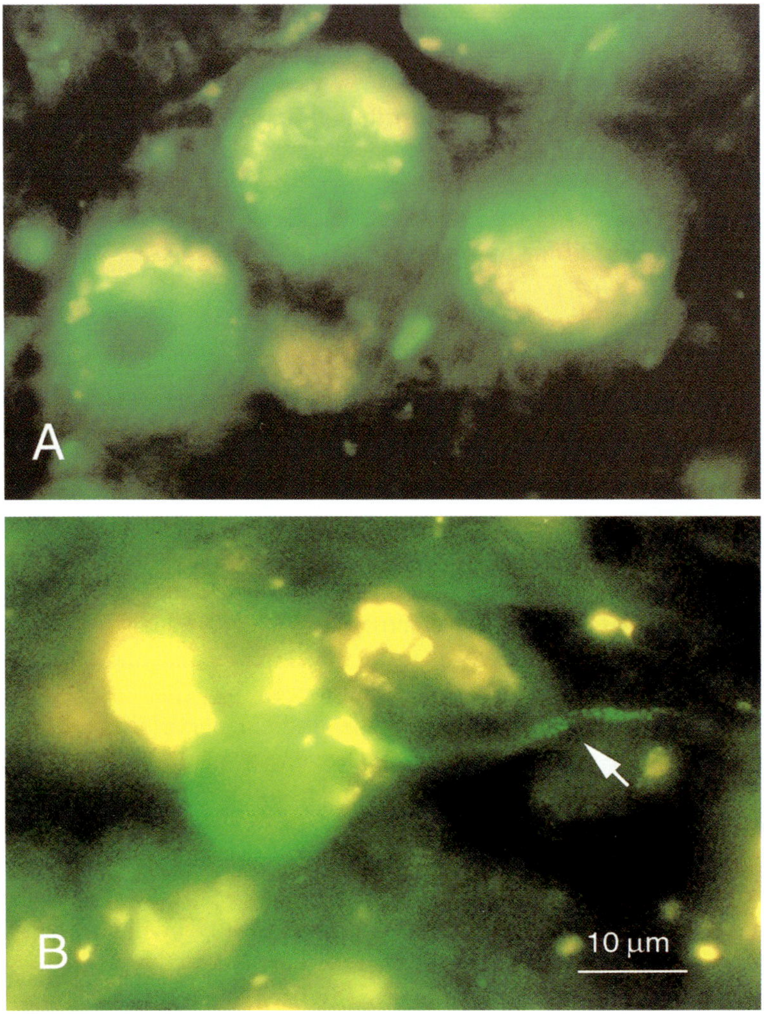

Fig. 1. SP-like immunoreactivity in human Scarpa's ganglion cells. Cryosections were immunostained with FITC to visualize SP. Specific SP staining is shown in green, whereas yellow-stained structures represent autofluorescence of lipofuscin granules. ***A*** SP-like immunofluorescence of three small ganglion cells. ***B*** A small SP-like labeled ganglion cell with SP-like granular immunoreactivity in the thin nerve process (arrow). The magnification (scale marker in ***B***) is identical in the two photomicrographs.

Fig. 2. Electrophysiological demonstrations of SP. ***a*** Repetitive intracellular stimulation of a vestibular ganglion cell before (control) and during the application of SP. ***b*** Perfusion of SP indicated by arrows evokes action potentials. Changes in membrane properties are illustrated below by current-voltage relationships. ***c*** Consecutive recording of fiber activity at the subsynaptic region of cochlear ICHs showing the effect of the SP antagonist spantide on neuronal firing induced by AMPA, NMDA and SP.

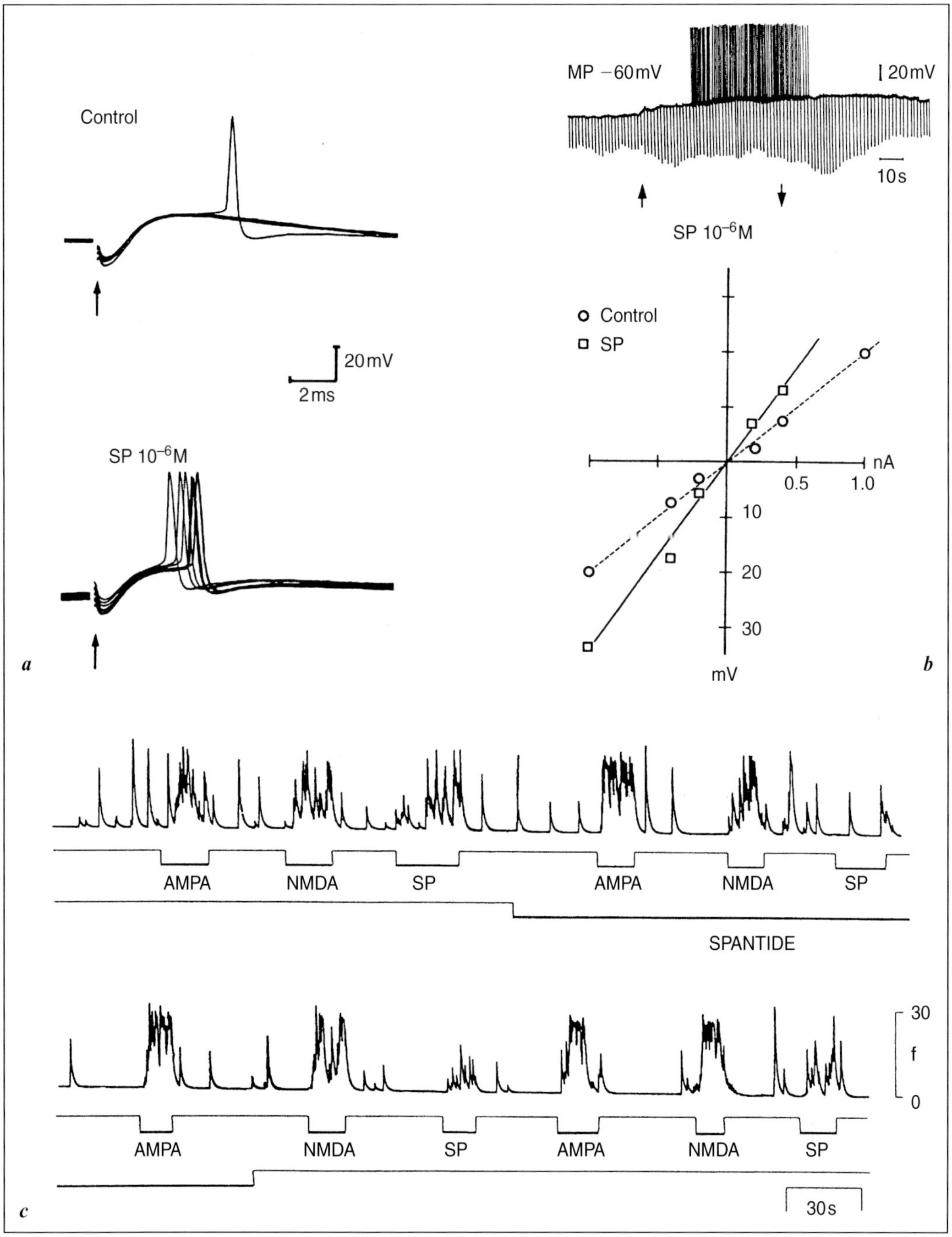
Control
20 mV
2 ms
SP 10⁻⁶M
a
MP −60 mV
20 mV
10 s
SP 10⁻⁶M
Control
SP
nA
0.5
1.0
10
20
30
mV
b
AMPA
NMDA
SP
AMPA
NMDA
SP
SPANTIDE
30
f
0
AMPA
NMDA
SP
AMPA
NMDA
SP
30 s
c

sensitization of human Scarpa's ganglion cells by SP may take place (fig. 2a, b). An intracellular study demonstrated slow depolarization by SP of the membrane potential which was accompanied by an increase in membrane resistance [11]. Furthermore, an enhanced firing in response to SP-induced depolarization occurred. Therefore, the vestibular peptidergic ganglion cells could act as the first relay station in which the incoming pulses are modulated. The findings strongly support the hypothesis that such a modulatory action is mediated by SP. Tsunoo et al. [12] have shown that SP acts as a slow excitatory transmitter of primary afferent neurons in guinea pig sympathetic ganglia. One may assume that SP-sensitive vestibular ganglion cells belong to the sensory system.

Although the observation of SP in the sensory neurons of the vestibular system could suggest that it might have a role as neurotransmitter released at central terminals, the distribution pattern of SP in the vestibular end-organs is somewhat intriguing. In the maculae SP was mainly found in the basal region of the sensory cells, whereas in the cristae SP immunoreactivity was strictly confined to the slopes [3, 9]. This is in agreement with the distribution pattern of SP-immunoreactive fibers predominantly innervating the slope of the cristae and the peripheral region of the maculae [6, 7]. Furthermore, the type of sensory cells linked to the presence of SP needs further elucidation. There are differences between mammalian and human vestibular sensory epithelia. Ylikoski et al. [5, 13] found in rabbits strong SP-like immunoreactivity in the calyx-shaped nerve terminals of type I sensory hair cells of the maculae in rabbits, although not consistently. Similar results were obtained in rats [14] and guinea pigs [6]. However, in humans, Felix et al. [9] showed that in addition to a small portion of calyces of type I hair cells, dot-like immunoreactivity occurred in bouton-like nerve endings of type II hair cells. The question arises whether an evolutionary aspect had to be taken into account besides a species difference.

The chemically distinct subpopulations of vestibular ganglion cells may be an indication of functional differences within vestibular afferent neurons. It would thus seem that a subpopulation of afferent neurons uses SP as a neurotransmitter. Furthermore, within the vestibular ganglion cell population SP could act as a facilitator. Interestingly, however, SP coexists with other neuropeptides in vestibular end-organs. SP and neurokinin-A were found to coexist in afferent fibers innervating the peripheral regions of both the utricular and ampullar sensory organs. Furthermore, calcitonin-gene-related peptide (CGRP) may also regulate the activity of the tachykinin-containing afferents [15]. Usami et al. [14] revealed three different types of SP-CGRP immunoreactivity in elements beneath and around vestibular hair cells, suggesting that different types of peripheral nervous systems may exist in the vestibular periphery.

The Cochlea

A modulator or neurotransmitter function for SP in the peripheral auditory system similar to that described in the vestibular organs has been hypothesized for a long time. After complete failure to demonstrate SP-like immunoreactivity in the organ of Corti, the spiral ganglion, the cochlear nerve and cochlear nuclei using a monoclonal antibody, Davis [16] concluded that SP had no function at all in the peripheral auditory system.

Despite the earlier failure or inaccurate results, the presence of SP in peripheral auditory systems is generally accepted today. There are few reports on the distribution of this peptide in the spiral ganglion. Ylikoski et al. [5] examined the presence of SP using an immunohistochemical technique and a monoclonal SP antibody. Fifty percent of the spiral ganglion cells innervating the basal part of the cochlea showed SP-like reactivity. Their findings were later confirmed by combined ultrastructural and histochemical methods. The presence and distribution of SP in a circumscribed spiral ganglion cell population as well as in fine fibers of the nerve trunk [13] suggests that the morphological basis of functional afferent nerve fibers is situated in this region. As described in the peripheral vestibular system, SP was identified in the spiral ganglion together with other neuroactive substances such as vasoactive intestinal peptide (VIP) tyrosine hydroxylase, CGRP and enkephalin [17].

SP seems to be widely distributed in the organ of Corti, especially in the inner (IHC) and outer hair cell (OHC) regions. It is of interest, however, that SP is not homogeneously distributed in both hair cell regions throughout the cochlea. The IHCs showed regular SP-like immunoreactivity in the basal part of the cells throughout all turns. There were IHCs where the SP positive staining appeared concentrated not only at the basal but also at the apical pole. As far as the number of labeled cells is concerned, however, an obvious increase was observed from the basal to the apical turn. In contrast to the IHCs, the OHCs showed a more diffuse SP-immunoreactive pattern. SP immunofluorescence was detected in the first row and to a minor degree in the second row. However, no labeling was found in the third row of the OHCs. As far as the distribution of immunoreactive cells in different turns is concerned, no labeling was observed in the basal turn. A slight increase was found from the second turn to the apex [18].

The distinct occurrence of SP immunoreactivity in neural elements beneath IHCs [13, 18–20] may be an indication of a functional relevance within the cochlea. This is further substantiated by the fact that during development the highest value of SP was found in the postnatal period [21]. The coexistence of SP with several different neuroactive substances, such as enkephalins, dynorphins, VIP, CGRP, neuropeptide Y, or dopamine as well as with excitatory amino acids acting as fast neurotransmitters suggests a multiplicity of neuromodulator or neurotransmitter systems in the cochlea.

Siegel and Dallos [22] demonstrated that spike activity can be recorded from dendrites underneath the IHCs. This unexpected finding led to experiments using the classical microiontophoretic technique on anesthetized animals. With this technique, different compounds could be tested and specific antagonistic substances could be administered in the immediate vicinity of the subsynaptic area of the IHC layer [23–25]. Ejection of SP resulted in a pronounced dose-dependent increase in subsynaptic spike activity similar to that observed following the application of the glutamate agonists N-methyl-*D*-aspartate (NMDA) and alpha-amino-3-hydroxy-5-methyl-4-isoxazolepropionic acid (AMPA). The selective SP antagonist spantide (*D*-Arg1, *D*-Trp7,9, Leu11-substance P) specifically blocked the SP-induced activity but without altering the activity of NMDA or AMPA [18]. The fact that the SP-induced activity is selectively blocked by the appropriate antagonist reveals different receptor sites for SP and glutamate in the cochlea (fig. 2c).

Although OHCs are more suitable for functional tests because they may be isolated and kept under physiological conditions for a long period, no clear picture about the role of SP on OHCs can be drawn. Whether there is a lack of functional significance or methodological failure needs further investigation. In his review, Eybalin [26] stated: 'The OHC afferent fiber synapse lacks a reliable indicator of activity… Thus, because conditions are rarely met in the electrophysiological studies reported and moreover would have significant influences on the responses of the IHCs and their afferent innervations, it must be considered that the data concerning putative OHC afferent neurotransmitter can only rely on morphological results.' In contrast to IHCs, where the action of SP seems to be excitatory [18], OHCs are hyperpolarized by SP by inducing outward currents in a dose-dependent manner. Using whole cell patch-clamp techniques, SP reduced nonselective cation channel conductance [27]. The authors suggested that the action of SP may not be mediated by the tachykinin receptor but rather by a tachykinin receptor independent pathway.

Conclusion

In recent years it has become clear that peptides not only have multiple effects, but are also broadly distributed throughout the brain and body. One may ask why SP should have so many functional roles and be so widely distributed. Because the peptide structure is small, it is easily formed and inactivated by enzyme action. Furthermore, the amino acid sequence offers a number of permutations. A large number of analogs can be produced which compete for receptor occupancy. The basic principle of action, however, seems identical: like any other neuropeptide, SP is mostly contained in neurons, released there upon excitation of the cell, and acts on receptors on target neurons to evoke various cellular responses. The actual receptors are the starting point for entry into the complex circuits that control vestibular or auditory functions. By defining these receptors, we should eventually be able to study these neural circuits for such physiological mechanisms in as precise a way as the sensory systems are being studied.

As shown in this review, SP is involved in both peripheral vestibular and auditory systems. Its distribution pattern together with its direct physiological action clearly demonstrates that this peptide acts as an excitatory neurotransmitter or facilitatory neuromodulator in regions where auditory or vestibular information processing takes place. To conclude, transmitter roles for SP in the cochlear and vestibular processes could provide new explanations for many physiological phenomena of hearing and vestibular function.

References

1 Otsuka M, Yoshioka K: Neurotransmitter functions of mammalian tachykinins. Physiol Rev 1993;73:229–308.

2 Holzer P: Capsaicin: cellular targets, mechanisms of action, and selectivity for thin sensory neurons. Pharmacol Rev 1991;43:143–201.

3 Demêmes D, Ryzhova I: Ontogenesis of substance P in the rat peripheral vestibular system. Hear Res 1997;114:252–258.

4 Felix H, Schickinger B, Felix D: Immunohistochemical and electrophysiological demonstration of substance P in human vestibular ganglion cells. Acta Otolaryngol (Stockh) 1996;116: 273–276.

5 Ylikoski J, Eranko L, Paivarinta H: Substance P-like immunoreactivity in the rabbit inner ear. J Laryngol Otol 1984;98:759–765.

6 Usami S, Hozawa J, Tazawa M, Jin H, Matsubara A, Fujita S: Localization of substance P-like immunoreactivity in guinea pig vestibular end-organs and the vestibular ganglion. Brain Res 1991;555:153–158.

7 Usami S, Hozawa J, Shinkawa H, Tazawa M, Jin H, Matsubara A, Fujita S, Ylikoski J: Immunocytochemical localization of substance P and neurofilament proteins in the guinea pig vestibular ganglion. Acta Otolaryngol (Stockh) 1993;503:127–131.

8 Carpenter MB, Huang Y, Pereira AB, Hersh LB: Immunocyto-chemical features of the vestibular nuclei in the monkey and cat. J Hirnforsch 1990;31:585–599.

9 Felix H, Johnsson LG, Gleeson MJ, Schickinger B, Felix D: Substance P in the human vestibular organ. 34th Workshop on Inner Ear Biology, Rosa Marina Italy 1997.

10 Felix H, Hoffmann V, Wright A, Gleeson M: Ultrastructure findings on human Scarpa's ganglion. Acta Otolaryngol (Stockh) Suppl 1987;436:85–92.

11 Felix D, Felix H: Intracellular study of substance P in human Scarpa's ganglion cells. ORL 1995;57:15–17.

12 Tsunoo A, Konishi S, Otsuka M: Substance P as an excitatory transmitter of primary afferent neurons in guinea-pig sympathetic ganglia. Neuroscience 1982;7:2025–2037.

13 Ylikoski J, Pirvola U, Happola O, Panula P, Virtanen I: Immunohistochemical demonstration of neuroactive substances in the inner ear of rat and guinea pig. Acta Otolaryngol (Stockh) 1989; 107:417–423.

14 Usami S, Hozawa J, Ylikoski J: Coexistence of substance P and calcitonin gene-related peptide-like immunoreactivities in the rat vestibular end-organs. Acta Otolaryngol (Stockh) 1991;481: 166–169.
15 Scarfone E, Ulfendahl M, Lundeberg T: The cellular localization of the neuropeptides substance P, neurokinin A, calcitonin gene-related peptide and neuropeptide Y in guinea-pig vestibular sensory organs: A high-resolution confocal microscopy study. Neuroscience 1996;75:587–600.
16 Davies WE: The absence of substance P from guinea pig auditory structures. Arch Otorhinolaryngol 1982;234:135–137.
17 Anniko M, Arnold W, Stigbrand T, Strom A: The human spiral ganglion. ORL 1995;57:68–77.
18 Schickinger B, Ehrenberger K, Felix D, Heiniger Burki C, Imboden H, Davies WE, Felix H: Substance P in the auditory hair cells in the guinea pig. ORL 1996;58:121–126.
19 Nowak R, Scholtz HJ, Rathsack R, Oehme P, Zelck U: Quantitative determination of substance P in the inner ear of the guinea pig. Pharmazie 1989;44:163–164.
20 Ulfendahl M, Lundeberg T, Scarfone E, Theodorsson E: Substance P in the guinea-pig hearing organ. Acta Physiol Scand 1993;148:357–358.
21 Nowak R, Zelck U, Rathsack R, Oehme P, Scholtz HJ, Koitschev A, Beleites B: Postnatal development of substance P in the inner ear of the guinea pig. Eur Arch Otorhinolaryngol 1990;247: 147–148.
22 Siegel JH, Dallos P: Spike activity recorded from the organ of Corti. Hear Res 1986;22:245–248.
23 Felix D, Ehrenberger K: The efferent modulation of mammalian inner hair cell afferents. Hear Res 1992;64:1–5.
24 Ehrenberger K, Felix D: Glutamate receptors in afferent cochlear neurotransmission in guinea pigs. Hear Res 1991;52:73–80.
25 Ehrenberger K, Felix D: Receptor pharmacological models for inner ear therapies with emphasis on glutamate receptors: A survey. Acta Otolaryngol (Stockh) 1995;115:236–240.
26 Eybalin M: Neurotransmitters and neuromodulators of the mammalian cochlea. Physiol Rev 1993;73:309–373.
27 Kakehata S, Akaike N, Takasaka T: Substance P decreases the non-selective cation channel conductance in dissociated outer hair cells of guinea pig cochlea. Ann NY Acad Sci 1993;707: 476–479.

Dr. Heidi Felix, ENT Department, University Hospital of Zürich,
Frauenklinikstrasse 24, CH–8091 Zürich (Switzerland)
Tel. +41 1 255 58 05, Fax +41 1 255 89 59, E-Mail felix@orl.usz.ch

Felix D, Oestreicher E (eds): Rational Pharmacotherapy of the Inner Ear.
Adv Otorhinolaryngol. Basel, Karger, 2002, vol 59, pp 35–41

Calcium Channels in Mouse Hair Cells: Function, Properties and Pharmacology

Jutta Engel[a], *Marcus Michna*[a], *Josef Platzer*[b], *Jörg Striessnig*[b]

[a] Institute of Physiology II and Department of Sensory Biophysics of the ENT Clinic, University of Tübingen, Germany;
[b] Institute for Biochemical Pharmacology, University of Innsbruck, Austria

Abstract

Adult inner hair cells (IHCs) possess voltage-activated Ca^{2+} currents that couple receptor potentials to transmitter release at the afferent synapses. Before the onset of hearing both IHCs and outer hair cells (OHCs) exhibit Ca^{2+} currents. More than 90% of neonatal hair cell (HC) currents flow through α1D Ca^{2+} channel subunits because they are absent in both IHCs and OHCs from $\alpha 1D^{-/-}$ mice and residual currents are insensitive to L-type agonists. Since lack of the α1D-subunit leads to HC degeneration and profound deafness, class D L-type Ca^{2+} currents seem to be crucial for the development and functioning of the inner ear. Neonatal HC Ca^{2+} currents were studied using the whole-cell patch clamp technique. They showed rapid activation, rapid deactivation and very little inactivation. They started activating as negative as -65 mV. In contrast to α1C-mediated (classical L-type) Ca^{2+} currents, they showed a rather low sensitivity to various L-type antagonists. 10 μM nifedipine e.g. blocked HC Ca^{2+} currents by about 40% whereas class C L-type Ca^{2+} currents are completely blocked by 100 nM nifedipine. The L-type channel agonist Bay K 8644 increased the HC Ca^{2+} current by 100–200% and shifted the IV curve to more negative potentials which is similar to its effects in α1C-mediated Ca^{2+} currents.

Function of Ca^{2+} Currents in Hair Cells

Ca^{2+} influx through voltage-activated Ca^{2+} channels serves various functions in both inner hair cells (IHCs) and outer hair cells (OHCs) which are summarized in table 1. In mature IHCs, Ca^{2+} channels provide the Ca^{2+} ions needed for coupling changes in the receptor potential to release glutamate [1]. At the same time Ca^{2+} ions that entered the cell activate large-conductance Ca^{2+}-dependent (BK) K^+ channels that guarantee for fast repolarization of the

Table 1. Established and presumptive functions of Ca^{2+} elevations and Ca^{2+} fluxes in immature and adult IHCs and OHCs of mice

	Before onset of hearing (<P12)	After onset of hearing
IHCs	slow Ca^{2+} action potentials [2, 3]	transmitter release [1]
	transmitter release [5] → forming functional connections in the auditory pathway	source of Ca^{2+} for BK Ca^{2+} activated K^+ channels → fast repolarization and fast time constant of IHC response [2]
	gene expression?	gene expression?
OHCs	slow evoked Ca^{2+} action potentials [4]?	transmitter release?
	transmitter release?	source of Ca^{2+} for BK (and SK?) Ca^{2+}activated K^+ channels? → fast repolarization
	gene expression?	gene expression?

receptor potential [2]. So far, it is not known if adult OHCs also have functional afferent synapses. Although ribbons, the specialized presynaptic bodies of hair cells (HCs), were found at the base of OHCs in many mammalian species, it is still a matter of debate whether type II afferents contacting OHCs are indeed functional.

Nevertheless, voltage-activated Ca^{2+} channels clearly play a role in the development of both IHCs and OHCs. Oscillations of the membrane potential have been recorded in acutely explanted IHCs of the neonatal mouse around postnatal day 6 (P6) i.e. well before the onset of hearing [2, 3]. In neonatal OHCs, which did not exhibit spontaneous oscillations, slow action potentials could be evoked by small current injections [4]. The slow time course of these action potentials (~50 ms duration) and their dependence on the extracellular Ca^{2+} concentration suggests that they are driven by the influx of Ca^{2+} through voltage-activated Ca^{2+} channels. The function of these Ca^{2+} action potentials in neonatal HCs might be: (1) generation of a pattern of activity that shapes functional synaptic connections in the auditory pathway [5], (2) Ca^{2+}-dependent expression of specific genes responsible for maturation of IHCs and OHCs.

Molecular Basis and Biophysical Properties of Ca^{2+} Currents in IHCs

Though it has been known for some time that (part) of the Ca^{2+} channels in mouse IHCs are of the L-type [6], the exact molecular nature of these channels has remained unknown. Voltage-activated Ca^{2+} channels consist of 4 (some even of 5) subunits: $\alpha1$, $\alpha2$–δ, β (and γ), with the $\alpha1$-subunit being the largest and most important subunit. $\alpha1$-subunits form the channel pore; they comprise the voltage sensor, the selectivity filter for Ca^{2+} ions, and the binding sites for all known toxins and drugs acting on voltage-activated Ca^{2+} channels [7]. L-type Ca^{2+} channels (LTCCs), which are characterized by a relatively high voltage of activation and by their sensitivity towards dihydropyridines (DHPs), are formed by 1 of the 4 $\alpha1$ genes $\alpha1C$, $\alpha1D$, $\alpha1F$ or $\alpha1S$. Recently a mouse mutant deficient for the $\alpha1D$-subunit ($\alpha1D^{-/-}$) has been constructed [8]. These mice are viable but show several defects such as congenital deafness. Anatomical inspection revealed that from P14 onwards OHCs started degenerating followed by IHCs and spiral ganglion cells [8]. Degeneration of HCs in $\alpha1D^{-/-}$ mice led to the conclusion that the development of functional IHCs and OHCs depends on functional class D LTCCs [8].

We investigated Ca^{2+} currents in HCs in wild-type ($\alpha1D^{+/+}$) and $\alpha1D^{-/-}$ mice by using the whole-cell configuration of the patch clamp technique. Pieces of the organ of Corti were used in which HC somata were exposed by sucking off the tectorial membrane and supporting cells with cleaning pipettes. The patch pipette solution contained Cs-gluconate to block most of the K^+ currents; the external solution contained Tris-Cl with 10 m*M* Ba^{2+} (or 10 m*M* Ca^{2+}) as charge carrier to maximise currents through Ca^{2+} channels and to block residual K^+ and Na^+ currents [8].

Figure 1 shows peak Ba^{2+} currents through Ca^{2+} channels in $\alpha1D^{+/+}$ (a, b) and $\alpha1D^{-/-}$ mice (c, d). Wild-type currents activated and deactivated rapidly, showed little inactivation (even over periods of more than 400 ms) and could be increased by 5 μM Bay K 8644, a DHP agonist, by a factor of 2–3 [8]. The mean current density for $\alpha1D^{+/+}$ IHCs from P3 to P9 was 46.8 $\pm$ 12.6 pA/pF (fig. 1e). IHCs from heterozygous mice ($\alpha1D^{+/-}$) had similar currents as those from wild-type IHCs whereas IHCs from $\alpha1D^{-/-}$ mice showed much smaller inward currents that were not affected by Bay K 8644 (fig. 1c, d). Their average current density was reduced to 3.7 $\pm$ 2.5 pA/pF. We conclude that about 92% of the total Ba^{2+} current in IHCs flows through class D ($\alpha1D$) LTCCs (fig. 1e).

Pharmacological Properties of Ca^{2+} Currents in IHCs

Application of different DHP antagonists such as nimodipine or nifedipine at rather high concentrations (10 μM) led to incomplete block of IHC Ba^{2+} currents: nimodipine decreased the current to 58 $\pm$ 8% (n = 4) of control [8], nifedipine to 56 $\pm$ 12% (n = 3). Blocking of the Ca^{2+} current was equally

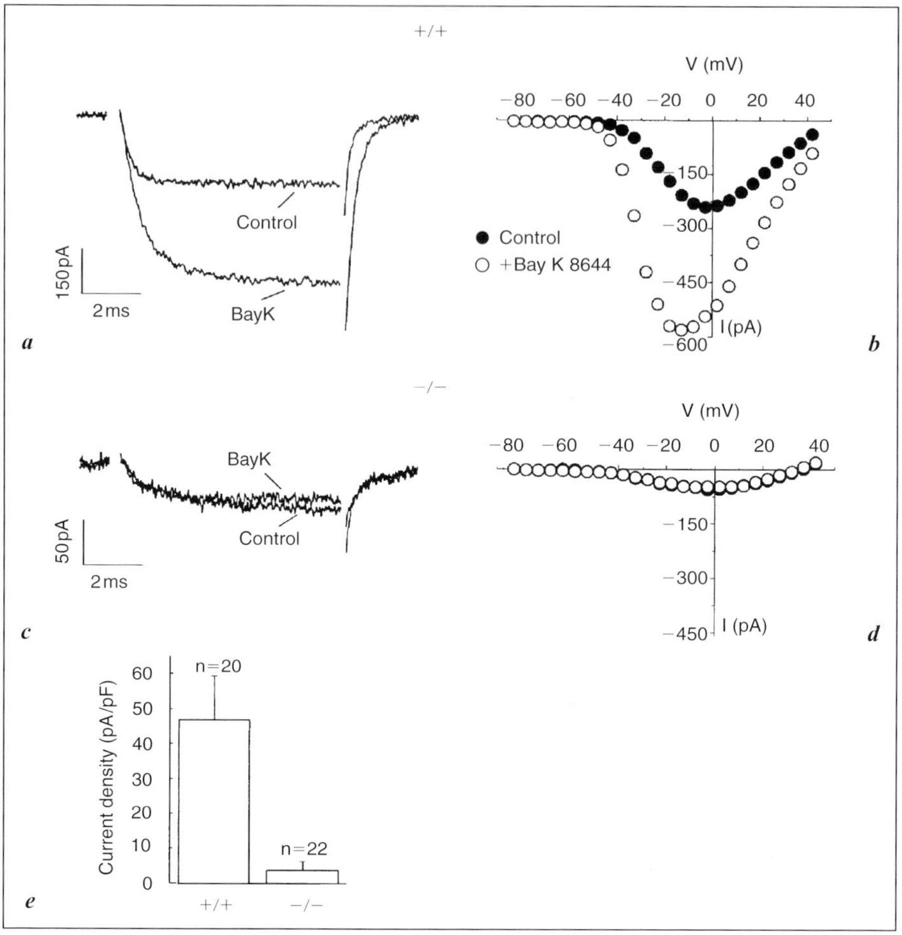

Fig. 1. Ba^{2+} currents through LTCCs in IHCs. Peak I_{Ba} current traces in IHCs isolated from 9-day-old $\alpha 1D^{+/+}$ (***a***) and $\alpha 1D^{-/-}$ mice (***c***) were elicited by 8-ms depolarizing steps from a holding potential of −83 mV to the potential of peak current (−2 mV, except −12 mV in ***a***, trace BayK), in the absence (control) and 2 min after superfusion with the DHP Ca^{2+} channel activator Bay K 8644 (BayK, 5 μ*M*). Corresponding IV curves (***b*** and ***d***) are also shown. ***e*** Mean current densities (±SD) for $\alpha 1D^{+/+}$ (+/+) and $\alpha 1D^{-/-}$ (−/−) obtained from the indicated number of cells. Reprinted from Platzer et al. [8], copyright 2000, with permission from Elsevier Science.

ineffective (fig. 2). The block was fully reversible, and the nifedipine-resistant current exhibited the same voltage characteristics as the control current shown by the scaled current-voltage (IV) curves (fig. 2c). Taking into account that >90% of total IHC current flows through class D LTCCs it must be concluded

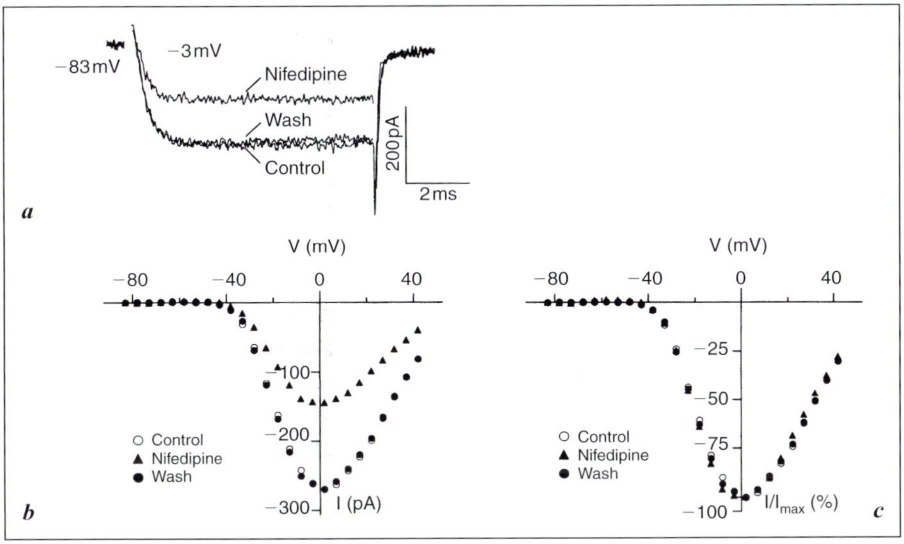

Fig. 2. Nifedipine blocks the IHC Ca^{2+} current reversibly but incompletely (P5, 10 m*M* Ca^{2+} as charge carrier). ***a*** Peak current traces before, during and after application of 10 μ*M* nifedipine. ***b*** Corresponding IV curves measured during the last millisecond of the depolarizing pulse. The three IV curves from ***b*** were scaled to their maximum (100%) to show that the residual current during nifedipine application does not differ in its voltage dependence from the current before/after DHP application (***c***).

that LTCCs are rather insensitive to blocking by nimodipine and nifedipine. For comparison, the IC_{50} for class C LTCCs is in the range of 10–100 n*M* such that 10 μ*M* DHP would completely block those channels [9]. Therefore patients treated for cardiovascular diseases at clinically relevant serum levels of DHPs should not suffer from a temporal threshold shift due to inhibition of the IHC synapse. On the other hand, DHPs should not be able to protect HCs from noise trauma, which has been demonstrated in an animal model (gerbils) [10]. So far, no specific blockers for class D LTCCs are known and no detailed studies exist on their pharmacology. This is due to the fact that (1) heterologous expression of α1D-subunits has not been successful so far and (2) there is no easily accessible mammalian cell type that expresses predominantly class D LTCCs.

Ca^{2+} Currents in Neonatal OHCs

Preliminary experiments suggest that neonatal OHCs exhibit Ca^{2+} currents that are very much like those of neonatal IHCs. Figure 3a and b shows

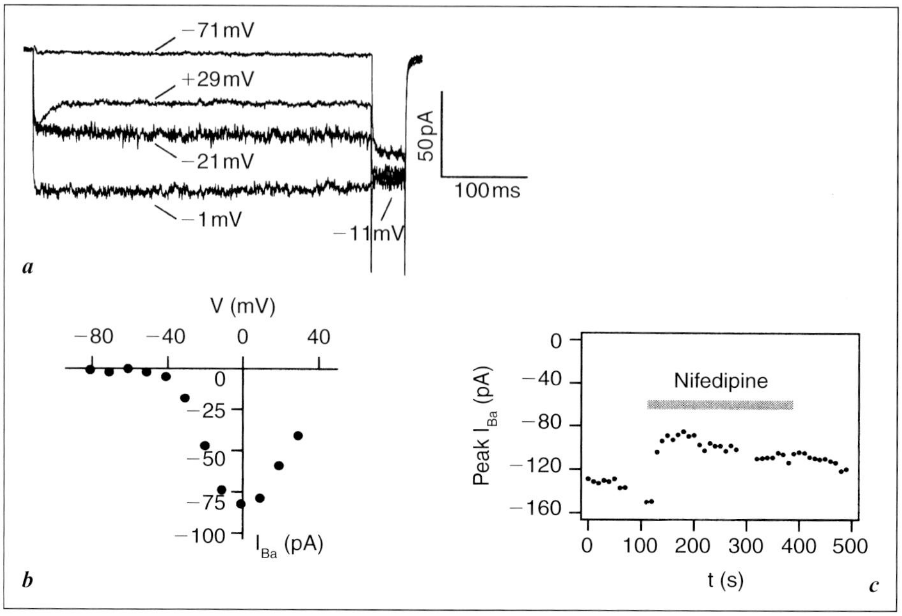

Fig. 3. Ba^{2+} currents in an OHC and action of nifedipine. ***a*** A 5-day-old OHC was depolarized from a holding potential of −91 mV to the potentials indicated which resulted in a fast-activating Ba^{2+} current that showed no inactivation within 400 ms unless the potential was stepped to values > −1 mV. ***b*** Corresponding IV curve taken between 10 and 15 ms after the start of the depolarizing step. The same cell was subjected to local perfusion with 10 μ*M* nifedipine (time of perfusion indicated by the grey bar) which resulted in a decline of the peak current amplitude (***c***).

OHC Ba^{2+} current responses to long-lasting depolarizing pulses and the corresponding IV curve. The main difference between OHCs and IHCs is the reduced average current density in OHCs (<50% of that of IHCs) that could be the reason for the lack of spontaneous action potentials in OHCs [4]. The nifedipine block was incomplete in OHCs (fig. 3c) as it was in IHCs. Therefore we suspect that a major part of the total Ca^{2+} current flows through class D LTCCs. To this end, we are currently studying Ca^{2+} currents in $\alpha 1D^{-/-}$ OHCs.

References

1 Moser T, Beutner D: Kinetics of exocytosis and endocytosis at the cochlear inner hair cell afferent synapse of the mouse. Proc Natl Acad Sci USA 2000;97:883–888.

2 Kros CJ, Ruppersberg JP, Rüsch A: Expression of a potassium current in inner hair cells during development of hearing in mice. Nature 1998;394:281–284.

3 Marcotti W, Géléoc GSG, Lennan GWT, Kros CJ: Transient expression of an inwardly rectifying potassium conductance in developing inner and outer hair cells along the mouse cochlea. Eur J Physiol 1999;439:113–122.
4 Marcotti W, Kros CJ: Developmental expression of the potassium current $I_{K,n}$ contributes to maturation of mouse outer hair cells. J Physiol 1999;520:653–660.
5 Beutner D, Ruesch A, Kros CJ, Moser T: Electrical activity of inner hair cells from the immature cochlea of mouse causes early stimulation of the auditory pathway. Assoc Res Otolaryngol Abstracts 2000;23:148.
6 Kros CJ, Rüsch A, Richardson GP, Russell IJ: Sodium and calcium currents in cultured cochlear hair cells of neonatal mice. J Physiol 1993;473:231P.
7 Hofmann F, Biel M, Flockerzi V: Molecular basis for Ca^{2+} channel diversity. Annu Rev Neurosci 1994;17:399–418.
8 Platzer J, Engel J, Schrott-Fischer A, Stephan K, Bova S, Chen H, Zheng H, Striessnig J: Congenital deafness and sinoatrial node dysfunction in mice lacking class D L-type Ca^{2+} channels. Cell 2000;102:89–97.
9 Morel N, Buryi V, Feron O, Gomez JP, Christen MO, Godfraind T: The action of calcium channel blockers on recombinant *L*-type calcium channel alpha1-subunits. Br J Pharmacol 1998;125: 1005–1012.
10 Boettcher FA, Caldwell RK, Gratton MA, White DR, Miles LR: Effects of nimodipine on noise-induced hearing loss. Hear Res 1998;121:139–146.

Note added in proof:
The human α1D subunit could be successfully expressed in heterologous systems recently, see: Bell DC, Butcher AJ, Berrow NS, Page KM, Brust PF, Nesterova A, Stauderman KA, Seabrook GR, Nurnberg B, Dolphin AC: Biophysical properties, pharmacology, and modulation of human, neuronal L-type (α_{1D}, $CA_v1.3$) voltage-dependent calcium currents. J Neurophysiol 2001;85: 816–827, and: Koschak A, Reimer D, Huber I, Grabner M, Glossmann H, Engel J, Striessnig J: α1D ($Ca_v1.3$) subunits form L-type Ca^{2+} channels activating at negative voltages. J. Biol. Chem. 2001;276:22100–22106.

Jutta Engel, Institute of Physiology II and Department of Sensory Biophysics of the ENT Clinic, University of Tübingen, Röntgenweg 11, D–72076 Tübingen (Germany)
Tel. +49 7071 2980806, Fax +49 7071 22917, E-Mail jutta.engel@uni-tuebingen.de

Felix D, Oestreicher E (eds): Rational Pharmacotherapy of the Inner Ear.
Adv Otorhinolaryngol. Basel, Karger, 2002, vol 59, pp 42–50

Adrenergic and Muscarinic Control of Cochlear Endolymph Production

Philine Wangemann

Anatomy and Physiology Department, Kansas State University, Manhattan, Kans., USA

Abstract

The transduction of sound into nerve impulses in the cochlea is dependent on the stria vascularis. It is a multilayered epithelium, which is part of the epithelial barrier between endolymph and perilymph. The current model designed to explain the generation of the endocochlear potential assumes that the molecular mechanism for the generation of the endocochlear potential is the $K_{IR}4.1$ K^+ channel localized in the intermediate cells and that strial marginal cells play an indirect role in the generation of the endocochlear potential. This role is limited to the maintenance of a low K^+ concentration in the intrastrial space by absorbing K^+ from this space and secreting it into the endolymph. The molecular mechanisms for K^+ secretion by strial marginal cells are well established. Strial marginal cells absorb K^+ from the intrastrial space via the Na^+-K^+-ATPase and the $Na^+2Cl^-K^+$cotransporter and secrete it across the apical membrane via the I_{sK}/K_vLQT1 K^+ channel. K^+ secretion by strial marginal cells is not only required for the maintenance of the endocochlear potential and to provide the charge carrier for the transduction mechanism, but also to maintain a constant volume of endolymph. Thus, the presence of multiple control mechanisms regulating the rate of K^+ secretion is likely. Recent observations suggest that the rate of K^+ secretion in strial marginal cells is stimulated by β_1-adrenergic receptors and inhibited by M_3 and/or M_4 muscarinic receptors.

Inner Ear Fluid Homeostasis

The endolymph has an unusual composition for an extracellular fluid in that the major salt is KCl rather than NaCl. Most other extracellular fluids including perilymph, cerebrospinal fluid and blood plasma contain NaCl as their major salt. The advantage of K^+ as the major cation in endolymph may rest

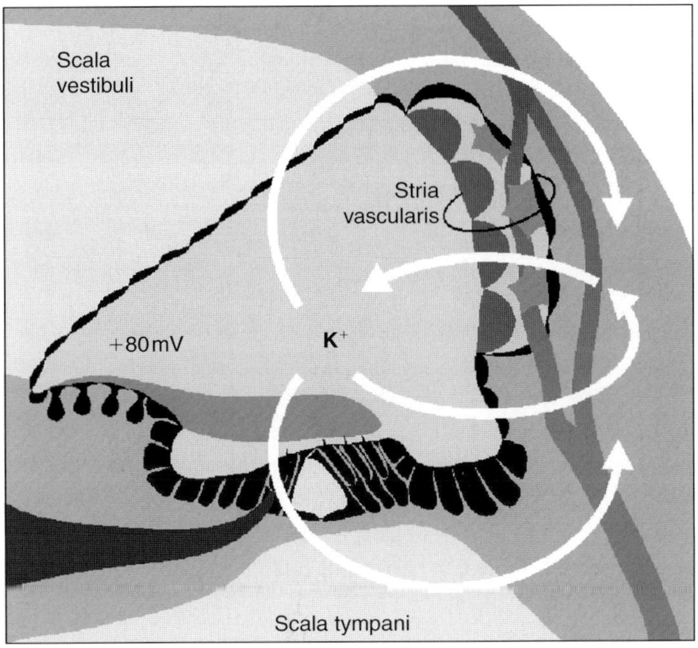

Fig. 1. Schematic cross-section through one turn of the cochlea. Note the multilayered structure of the stria vascularis.

with the necessity that the enclosing epithelial cells must carefully control their own cell volume as well as the fluid volume of endolymph. The control of cell volume and of endolymph volume is necessary to prevent mechanical disturbances incompatible with the mechanosensory function of the inner ear. Improper control of endolymph volume may result in a swelling of the endolymphatic compartment, a situation termed endolymphatic hydrops or, alternatively, may result in a collapse of the endolymphatic compartment. The elucidation of mechanisms maintaining inner ear fluid homeostasis may provide grounds on which to base strategies for the rational treatment of Ménière's disease.

Stria vascularis

The stria vascularis is a multilayered epithelium, which is part of the epithelial barrier between endolymph and perilymph (fig. 1). It faces on the apical side the endolymph and on the basal side a perilymph-like fluid that fills the extracellular spaces between the fibrocytes of the spiral ligament. The stria vascularis provides two electrochemical barriers consisting of epithelial cells

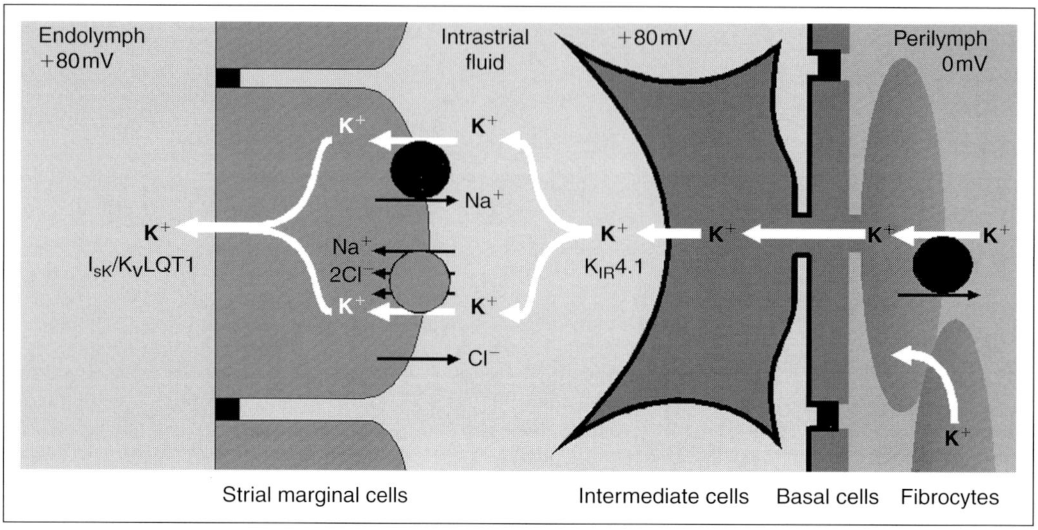

Fig. 2. Schematic cross-section through the stria vascularis. Note that it consists of two barriers, the marginal cell barrier and the basal cell barrier.

joined together by tight junctions [1]. The barrier between endolymph and the intrastrial fluid is comprised of strial marginal cells and the barrier between the intrastrial fluid and perilymph is comprised of basal cells. Further, the barrier between intrastrial fluid and plasma is comprised of endothelial cells. Strial marginal cells are unusual in that they are neither connected via gap junctions among each other nor to other cells of the stria vascularis [2]. In contrast, basal cells are connected via gap junctions to intermediate cells located inside the stria vascularis and to pericytes surrounding the capillary network in the center of the stria vascularis as well as to fibrocytes of the spiral ligament [3].

Stria vascularis Secretes K^+

The unusual composition of the endolymph is the result of ion transport processes, which take place in the enclosing epithelial cells. K^+ is secreted into the cochlear endolymph by strial marginal cells [4, 5]. K^+ and Na^+ are reabsorbed from endolymph by the sensory hair cells and by outer sulcus cells [6–8]. This chapter focuses on K^+ secretion and selected aspects of its regulation.

Strial marginal cells take up K^+ across the basolateral membrane via the Na^+-K^+-ATPase and the $Na^+2Cl^-K^+$ cotransporter and secrete K^+ across the apical membrane via the I_{sK}/K_vLQT1 K^+ channel (fig. 2). The presence and

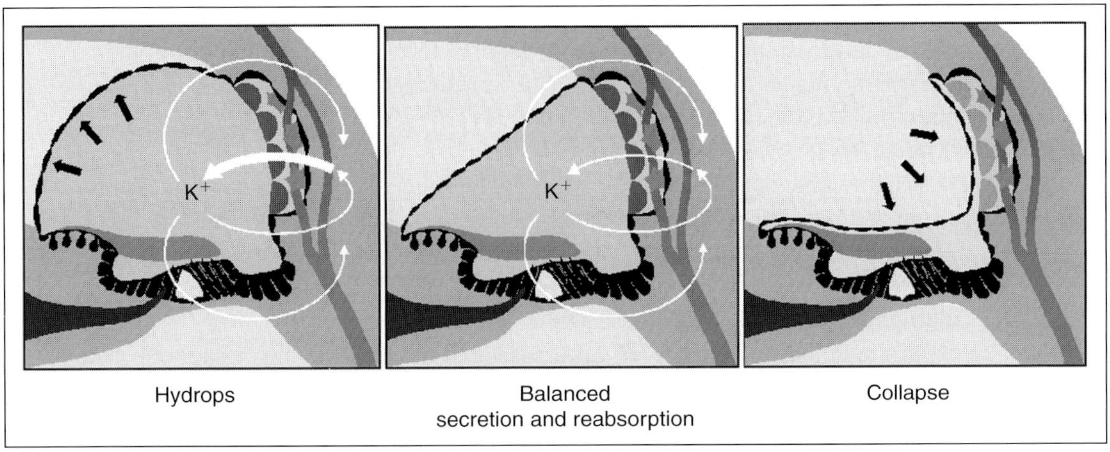

Fig. 3. Schematic cross-sections through one turn of the cochlea under conditions when K^+ secretion and reabsorption are balanced (middle), when K^+ secretion outweighs reabsorption (left) and when reabsorption outweighs K^+ secretion (right).

function of these ion transport mechanisms are now well established. The Na^+-K^+-ATPase, its presence and its role in ion transport have been demonstrated [4, 9]. Immunohistochemical, functional and molecular biological data have been obtained for the $Na^+2Cl^-K^+$ cotransporter [4, 10–12]. The presence and function of the I_{sK}/K_vLQT1 K^+ channel has been demonstrated in immunohistochemical, functional and molecular studies [4, 13–17]. Na^+ and Cl^- taken up via the $Na^+2Cl^-K^+$ cotransporter are recycled across the basolateral membrane via the Na^+-K^+-ATPase and Cl^- channels, respectively. The presence and function of the Cl^- channels are now well established [4, 18, 19]. Consistent with the localization of K^+ channels in the apical membrane and Cl^- channels in the basolateral membrane is the finding that the apical membrane is primarily K^+ conductive and the basolateral membrane is primarily Cl^- conductive [20]. The I_{sK} and the K_vLQT1 subunit of the I_{sK}/K_vLQT1 K^+ channel and the $Na^+2Cl^-K^+$ cotransporter have each been found to be essential for K^+ secretion and consequently for the formation of endolymph and the survival of the sensory hair cells. Strial marginal cells of mice lacking either one of the subunits of the I_{sK}/K_vLQT1 K^+ channel or lacking the $Na^+2Cl^-K^+$ cotransporter have been shown to be unable to secrete K^+ or to generate endolymph [12, 17, 21, 22]. The scala media of the cochlea appears collapsed in these mice (fig. 3). Apparently, reabsorptive processes continue to function when K^+ secretion is disabled. The collapsed endolymphatic compartment of mice lacking a functional

I_{sK}/K_vLQT1 K^+ channel resembles a human condition, the Jervell and Lange-Nielsen cardioauditory syndrome, a rare disorder in which congenital bilateral deafness coincides with a pathologically prolonged Q–T interval of the heart action potential leading to life-threatening cardiac arrhythmia. The connection between the inner ear and the heart is the I_{sK}/K_vLQT1 K^+ channel that is an essential mechanism for K^+ secretion in the inner ear and plays an accessory role in the repolarization phase of the cardiac action potential. Indeed, it has recently been found that 12–50% of patients with long-Q–T syndrome have a mutation in the K_vLQT1 subunit of the I_{sK}/K_vLQT1 K^+ channel [23, 24]. The finding that the I_{sK}/K_vLQT1 K^+ channel is the sole mechanism for K^+ secretion across the apical membrane of strial marginal cells makes this channel a fine pharmacologic target. Inhibition of the I_{sK}/K_vLQT1 K^+ channel could be of therapeutic value for the treatment of endolymph hydrops, a condition where endolymph secretion outweighs reabsorption resulting in a pathologic swelling of the endolymphatic compartment. A pharmacologic reduction of K^+ secretion could restore the balance between K^+ secretion and reabsorption and thus alleviate endolymphatic hydrops. Research is warranted to develop otoselective I_{sK}/K_vLQT1 K^+ channel blockers.

Stria vascularis Generates the Endocochlear Potential

The transduction of the mechanical sound stimulus into an electrical signal takes place in the stereocilia of the hair cells. The transduction current is carried by K^+ and driven by the sum of the membrane voltage of -40 to -70 mV generated by the basolateral membrane of the hair cells and the endocochlear potential of $+80$ mV. The endocochlear potential is a K^+ equilibrium potential generated by the low K^+ concentration in the intrastrial fluid, the high intracellular K^+ concentration of the intermediate cells and the $K_{IR}4.1$ K^+ channel in the membranes of the intermediate cells [25]. Intermediate cells are connected via gap junctions to the basal cells of the stria vascularis so that the membranes of the intermediate cells are physiologically connected to the inner membranes of the basal cells. Basal cells, which are connected among each other with tight junctions, form the barrier across which the endocochlear potential can be measured [26]. Evidence for an involvement of intermediate cells in the generation of the endocochlear potential comes from the observation that the endocochlear potential is absent in mice lacking intermediate cells [27]. The concept that the $K_{IR}4.1$ K^+ channel in the intermediate cells generates the endocochlear potential is supported by pharmacologic data showing that the endocochlear potential and the $K_{IR}4.1$ channel in intermediate cells share a pattern of sensitivities to various K^+ channel blockers [28, 29].

Regulation of K^+ Secretion

Endolymph volume homeostasis encompasses the control of K^+ secretion into endolymph as well as the control of K^+ reabsorption from endolymph. Ongoing research has identified a number of mechanisms, which control K^+ secretion in strial marginal cells. It is apparent that these epithelial cells integrate local and systemic signals including the presence of adrenergic [30], muscarinic [31, 32] and purinergic agonists [16, 33]. The current understanding of the regulation by β_1-adrenergic and by muscarinic receptors is briefly reviewed here.

β_3-Adrenergic Receptors Stimulate K^+ Secretion

Strial marginal cells contain β_1-adrenergic receptors in their basolateral membrane, which are linked to G_s protein and adenylyl cyclase and increase the rate of K^+ secretion via an elevation of cAMP and stimulation of the I_{sK}/K_vLQT1 K^+ channel [30]. Pharmacologic data unambiguously support the presence and location, the involvement of cAMP and the link to K^+ secretion [30]. G_s protein and adenylyl cyclase have been localized to the basolateral membrane of strial marginal cells [34, 35] and stimulation of adenylyl cyclase has been shown to incease the rate of K^+ secretion via stimulation of the I_{sK}/K_vLQT1 K^+ channel [36]. Under physiologic conditions the β_1-adrenergic pathway may stimulate K^+ secretion to match challenges during states of general excitation. Constantly elevated norepinephrine concentrations, such as found in the blood plasma of patients suffering from Ménière's disease, however, may result in an overstimulation of K^+ secretion and contribute to the formation of an endolymphatic hydrops. Interestingly, the β-adrenergic receptor antagonist propanolol has recently been shown to reduce the magnitude of endolymphatic hydrops in the Kimura animal model of endolymphatic hydrops [37].

M_3 and/or M_4 Muscarinic Receptors Inhibit K^+ Secretion

Strial marginal cells contain M_3 and/or M_4 muscarinic acetylcholine receptors in their basolateral membrane, which via an increase in the cytosolic Ca^{2+} concentration and a decrease in the cytosolic cAMP concentration cause an inhibition of K^+ secretion [31, 32; E.Q. Scherer, M. Herzog and P. Wangemann, unpubl. observations]. The question regarding the physiologic relevance of the muscarinic acetylcholine receptors in the stria vascularis is linked to the question of whether or not these receptors are reached by agonists. Interestingly, acetylcholine is the main efferent neurotransmitter, which exerts an inhibitory effect on the sensory cells and their innervation to protect them from overstimulation [38]. Although the stria vascularis is thought not to receive any direct innervation, it is conceivable that efferent neurotransmitters may reach strial marginal cells via the blood stream to exert an inhibitory and potentially protective effect. This concept, however, is as of now highly speculative.

Conclusion

K^+ secretion in the stria vascularis is under tight control by G-protein-coupled receptors including β_1-adrenergic and M_3 and/or M_4 muscarinic acetylcholine receptors. It is conceivable that over-stimulation of K^+ secretion via β_1-adrenergic receptors can lead to endolymphatic hydrops and conversely that overinhibition of K^+ secretion can lead to collapse of the endolymphatic compartment as observed in mice lacking ion transport mechanisms crucial for K^+ secretion (fig. 3). Ion transporters involved in K^+ secretion and the mechanisms involved in their regulation may provide suitable drug targets for the pharmacologic control of endolymph volume and the treatment of Ménière's disease.

Acknowledgements

The support by Research Grant R01-DC-01098 from the National Institute on Deafness and Other Communication Disorders, National Institutes of Health, is gratefully acknowledged.

References

1 Jahnke K: The fine structure of freeze-fractured intercellular junctions in the guinea pig inner ear. Acta Otolaryngol Suppl (Stockh) 1975;336:1–40.
2 Kikuchi T, Kimura RS, Paul DL, Takasaka T, Adams JC: Gap junction systems in the mammalian cochlea. Brain Res Brain Res Rev 2000;32:163–166.
3 Takeuchi S, Ando M: Dye-coupling of melanocytes with endothelial cells and pericytes in the cochlea of gerbils. Cell Tissue Res 1998;293:271–275.
4 Wangemann P, Liu J, Marcus DC: Ion transport mechanisms responsible for K^+ secretion and the transepithelial voltage across marginal cells of stria vascularis in vitro. Hear Res 1995;84:19–29.
5 Wangemann P: Comparison of ion transport mechanisms between vestibular dark cells and strial marginal cells. Hear Res 1995;90:149–157.
6 Johnstone BM, Patuzzi R, Syka J, Sykova E: Stimulus-related potassium changes in the organ of Corti of guinea-pig. J Physiol (Lond) 1989;408:77–92.
7 Marcus DC, Chiba T: K^+ and Na^+ absorption by outer sulcus epithelial cells. Hear Res 1999;134:48–56.
8 Chiba T, Marcus DC: Nonselective cation and BK channels in apical membrane of outer sulcus epithelial cells. J Membr Biol 2000;174:167–179.
9 Schulte BA, Adams JC: Distribution of immunoreactive Na^+, K^+-ATPase in gerbil cochlea. J Histochem Cytochem 1989;37:127–134.

10 Crouch JJ, Sakaguchi N, Lytle C, Schulte BA: Immunohistochemical localization of the Na-K-Cl co-transporter (NKCC1) in the gerbil inner ear. J Histochem Cytochem 1997;45:773–778.

11 Delpire E, Lu J, England R, Dull C, Thorne T: Deafness and imbalance associated with inactivation of the secretory Na-K-2Cl co-transporter. Nat Genet 1999;22:192–195.

12 Flagella M, Clarke LL, Miller ML, Erway LC, Giannella RA, Andringa A, et al: Mice lacking the basolateral Na-K-2Cl cotransporter have impaired epithelial chloride secretion and are profoundly deaf. J Biol Chem 1999;274:26946–26955.

13 Sakagami M, Fukazawa K, Matsunaga T, Fujita H, Mori N, Takumi T, et al: Cellular localization of rat Isk protein in the stria vascularis by immunohistochemical observation. Hear Res 1991;56: 168–172.

14 Neyroud N, Tesson F, Denjoy I, Leibovici M, Donger C, Barhanin J, et al: A novel mutation in the potassium channel gene K_VLQT1 causes the Jervell and Lange-Nielsen cardioauditory syndrome. Nat Genet 1997;15:186–189.

15 Shen Z, Marcus DC, Sunose H, Chiba T, Wangemann P: I_{sK} channel in strial marginal cell: Voltage-dependence, ion selectivity, inhibition by 293B and sensitivity to clofilium. Audit Neurosci 1997;3:215–230.

16 Marcus DC, Sunose H, Liu J, Bennett T, Shen Z, Scofield MA, et al: Protein kinase C mediates P_{2U} purinergic receptor inhibition of K^+ channel in apical membrane of strial marginal cells. Hear Res 1998;115:82–92.

17 Vetter DE, Mann JR, Wangemann P, Liu Z, McLaughlin KJ, Lesage F, et al: Inner ear defects induced by null mutation of *isk* gene. Neuron 1996;17:1251–1264.

18 Takeuchi S, Irimajiri A: $Cl^-$and nonselective cation channels in the basolateral membrane of strial marginal cells. Proc Sendai Symp 1994;4:35–37

19 Ando M, Takeuchi S: mRNA encoding ClC-K1, a kidney Cl^- channel, is expressed in marginal cells of the stria vascularis of rat cochlea: Its possible contribution to Cl^- currents. Neurosci Lett 2000;284:171–174.

20 Wangemann P, Marcus DC: The membrane potential of vestibular dark cells is controlled by a large Cl^-conductance. Hear Res 1992;62:149–156.

21 Dixon MJ, Gazzard J, Chaudhry SS, Sampson N, Schulte BA, Steel KP: Mutation of the Na-K-Cl co-transporter gene Slc12a2 results in deafness in mice. Hum Mol Genet 1999;8:1579–1584.

22 Francis H, Lustig L, Gorelikow M, Limb C, Lee M, Feinberg A: Otopathology of the KvLQT1 knockout mouse: A new murine model for deafness. Assoc Res Otolaryngol 2000;23: 419–419.

23 Tanaka T, Nagai R, Tomoike H, Takata S, Yano K, Yabuta K, et al: Four novel K_VLQT1 and four novel HERG mutations in familial long-QT syndrome. Circulation 1997;95:565–567.

24 Wang Q, Curran ME, Splawski I, Burn TC, Millholland JM, VanRaay TJ, et al: Positional cloning of a novel potassium channel gene: K_VLQT1 mutations cause cardiac arrhythmias. Nat Genet 1996;12:17–23.

25 Ando M, Takeuchi S: Immunological identification of an inward rectifier K^+ channel ($K_{ir}4.1$) in the intermediate cell (melanocyte) of the cochlear stria vascularis of gerbils and rats. Cell Tissue Res 1999;298:179–183.

26 Salt AN, Melichar I, Thalmann R: Mechanisms of endocochlear potential generation by stria vascularis. Laryngoscope 1987;97:984–991.

27 Carlisle L, Steel K, Forge A: Endocochlear potential generation is associated with intercellular communication in the stria vascularis: Structural analysis in the viable dominant spotting mouse mutant. Cell Tissue Res 1990;262:329–337.

28 Takeuchi S, Ando M: Inwardly rectifying K^+ currents in intermediate cells in the cochlea of gerbils: A possible contribution to the endocochlear potential. Neurosci Lett 1998;247:175–178.

29 Takeuchi S, Kakigi A, Takeda T, Saito H, Irimajiri A: Intravascularly applied K^+-channel blockers suppress differently the positive endocochlear potential maintained by vascular perfusion. Hear Res 1996;101:181–185.

30 Wangemann P, Liu J, Shimozono M, Schimanski S, Scofield MA: K^+ secretion in strial marginal cells is stimulated via β_1-adrenergic receptors but not via β_2-adrenergic or vasopressin receptors. J Membr Biol 2000;175:191–202.

31 Shimozono M, Scofield MA, Wangemann P: Molecular evidence for muscarinic receptor subtypes in stria vascularis and vestibular labyrinth. Assoc Res Otolaryngol. 1998;22.

32 Liu J, Wangemann P: Pharmacological evidence for the M_3 muscarinic receptor mediating inhibition of K^+ secretion in vestibular dark cells and strial marginal cells. Assoc Res Otolaryngol. 1998;21.

33 Liu J, Kozakura K, Marcus DC: Evidence for purinergic receptors in vestibular dark cell and strial marginal cell epithelia of the gerbil. Audit Neurosci 1995;1:331–340.

34 Zajic G, Anniko M, Schacht J: Cellular localization of adenylate cyclase in the developing and mature inner ear of the mouse. Hear Res 1983;10:249–261.

35 Mizuta K, Iwasa KH, Simonds WF, Tachibana M: Ultrastructural localization of G-protein G_s in the lateral wall of the guinea pig cochlear duct. Hear Res 1996;93:111–119.

36 Sunose H, Liu J, Marcus DC: cAMP increases K^+ secretion via activation of apical I_{sK}/ K_vLQT1 channels in strial marginal cells. Hear Res 1997;114:107–116.

37 Kimura RS: An update of our experiences on endolymphatic hydrops. Proceedings of the 4th International Symposium on Ménière's Disease, Paris, France. The Hague, Kugler, 2000.

38 Guinan JJ: Physiology of olivocochlear efferents; in Dallos P, Popper AN, Fay R (eds): The Cochlea. New York, Springer, 1996, pp 435–502.

Dr. Philine Wangemann, Associate Professor, Anatomy and Physiology Department,
Kansas State University, 1600 Dension Avenue, Manhattan, KS 66506 (USA)
Tel. +1 785 532 4863, Fax +1 785 532 4557, E-Mail wange@vet.ksu.edu

Felix D, Oestreicher E (eds): Rational Pharmacotherapy of the Inner Ear.
Adv Otorhinolaryngol. Basel, Karger, 2002, vol 59, pp 51–57

Cochlear Blood Flow Regulation

Philine Wangemann

Anatomy and Physiology Department, Kansas State University,
Manhattan, Kans., USA

Abstract

The regulation of cochlear blood flow is crucial for auditory function due to the sensitivity of this sensory organ to hypoxia. Part of the regulation of cochlear blood flow occurs in the spiral modiolar artery, which provides the main blood supply to the cochlea. Blood flow in general is most effectively regulated through the control of the vascular diameter. The vascular diameter is determined by the degree of constriction of the smooth muscle cells in the vascular wall. A constriction of the smooth muscle cells reduces the diameter of the vascular lumen and thereby decreases blood flow, whereas a relaxation of the smooth muscle cells increases blood flow. The degree of constriction of the smooth muscle cells in the spiral modiolar artery is carefully controlled and must be adjusted properly to the demands of the cochlear tissues. To achieve proper control, smooth muscle cells integrate information from various sources. Vasoconstrictors and dilators may originate from the innervation surrounding the vessel, from endothelial cells lining the vascular lumen or from the smooth muscle cells themselves. Recent advances revealed that smooth muscle cells from different arterioles differ widely in their endowment with mechanisms that regulate the degree of smooth muscle cell tone. Signal transduction mechanisms, which mediate these neurogenic, local and paracrine regulations of smooth muscle contractility are now beginning to be understood. This report reviews recently obtained evidence for adrenergic regulation of cochlear blood flow and then focuses on a novel vasodilation mechanism that involves ryanodine receptors, Ca^{2+} sparks and the activation of Ca^{2+}-activated K^+ channels.

The field of vascular physiology has identified a number of pathways leading to vasoconstriction or vasodilation. Recent advances revealed that vascular smooth muscle cells of different origin differ widely in their endowment with mechanisms that regulate vascular tone. Thus, it is not possible to assume that mechanisms, which control vascular diameter in one blood vessel, exist in all vessels including the spiral modiolar artery. Detailed investigations of the spiral modiolar artery are therefore warranted.

Investigations of the spiral modiolar artery are aided by recently developed in vitro preparations [1–3]. These preparations in concert with video microscopy, microfluorometry, confocal microscopy and biochemical and molecular biological techniques promise to lead to significant advances in the understanding of cochlear blood flow regulation.

Adrenergic Regulation of Cochlear Blood Flow

Cochlear blood flow regulation via the sympathetic nervous system and adrenergic receptors is now well established. Sympathetic nerve fibers originating from the superior cervical and stellate ganglia have been shown to innervate the spiral modiolar artery [4–6]. This innervation contributes to the basal vascular tone of the cochlear vasculature. Indeed, blockade of the innervation originating from the stellate ganglion has been shown to release the vascular tone and increase cochlear blood flow [7]. Consistently, activation of the sympathetic nerve fibers by stimulation of the stellate ganglion or the superior cervical ganglion has been shown to reduce cochlear blood flow [8–10]. This reduction in cochlear blood flow was found to be attenuated by α-adrenergic receptor antagonists, which were applied at the round window [8].

Data obtained in situ provided evidence for an involvement of α_1- and α_2-adrenergic receptors in the sympathetic regulation of cochlear blood flow. Evidence suggesting an involvement of α_1-adrenergic but not α_2-adrenergic receptors is based on measurements of cochlear blood flow in conjunction with the application of agonists and antagonists at the round window [11]. The presence of α_1-adrenergic receptors is further supported by the finding of a high-affinity binding site for prazosin in studies involving whole cochleae [12]. In contrast, evidence suggesting a predominance of α_2-adrenergic receptors mediating regulation of cochlear blood flow is based on intravital diameter measurements of radiating arterioles in conjunction with infusions of agonists and antagonists [13]. A role of α_2-adrenergic receptors is further supported by the observation that an α_2-adrenergic receptor antagonist attenuated an increase in cochlear blood flow that was induced by electric field stimulation of the entire cochlea [14]. It is most conceivable that α_1- and α_2-adrenergic receptors are both involved in the regulation of cochlear blood flow and that these receptors are differentially distributed in the vasculature of the cochlea and that species differences contributed to the differences observed.

Recent data obtained in the isolated spiral modiolar artery have confirmed evidence for a sympathetic control of the vascular diameter. A vasoconstrictor can indeed be released from the neural network surrounding the spiral modiolar artery [15, 16]. Further, the smooth muscle cells, especially those at the four branch points of the gerbil spiral modiolar artery, contain α_{1A}-adrenergic receptors and respond to the α-adrenergic agonist norepinephrine with a vasoconstriction [17]. The observation that functional α_{1A}-adrenergic receptors are concentrated at the branch points, where radiating arterioles branch off from the spiral modiolar artery proper, is of great interest. It is conceivable that a constriction of these smooth muscle cells changes the angle between the radiating arteriole and the spiral modiolar artery. A change in the angle at the branch point would alter the distribution of the cellular fraction of blood such that an increase in the angle would reduce the cellular fraction distributed to the branch whereas a decrease in the angle would cause a more even distribution [18]. A controlled distribution of the cellular fraction of the perfused blood may alter the potential for oxygenation of the tissues supplied. Since the arteriolar branches of the spiral modiolar artery supply different regions of the cochlea, it is conceivable that the sympathetic control of the branch points contributes to a tonotopic regulation of cochlear blood flow. This concept, however, is currently highly speculative and remains to be experimentally proven.

Ryanodine Receptors, Ca^{2+} Sparks, Ca^{2+}-Activated K^+ Channels and Voltage-Gated L-Type Ca^{2+} Channels

Ca^{2+} is a very versatile signaling molecule in the spiral modiolar artery. The extracellular Ca^{2+} concentration sets the vascular tone of the spiral modiolar artery via L-type voltage-gated Ca^{2+} channels [19] and elevations of the extracellular Ca^{2+} concentration mediate a vasoconstriction via G-protein-coupled calcium-sensing receptors [20]. Elevations of the cytosolic Ca^{2+} concentration can be part of a Ca^{2+} wave or part of a local Ca^{2+} spark [21]. Ca^{2+} waves are transient increases in the cytosolic Ca^{2+} concentration, which last for 3–6 s, encompass the entire cytosol of the cell and cause a vasoconstriction (fig. 1). In contrast, Ca^{2+} sparks are transient increases in the cytosolic Ca^{2+} concentration, which last only 50–150 ms and remain localized in the space between the plasma membrane and the underlying sarcoplasmic reticulum (fig. 1). Ca^{2+} sparks in smooth muscle cells of other vessels, skeletal muscle or cardiac muscle are known to be due to transient openings of ryanodine receptors located in the sarcoplasmic reticulum. Ca^{2+} released via ryanodine receptors in form of a Ca^{2+} spark does not reach the contractile filaments and thus does not cause a vasoconstriction. Ca^{2+} sparks in vascular smooth muscle cells rather cause a vasodilation through opening of Ca^{2+}-activated K^+ channels, a hyperpolarization of the plasma membrane and consequent closure of the L-type voltage-gated

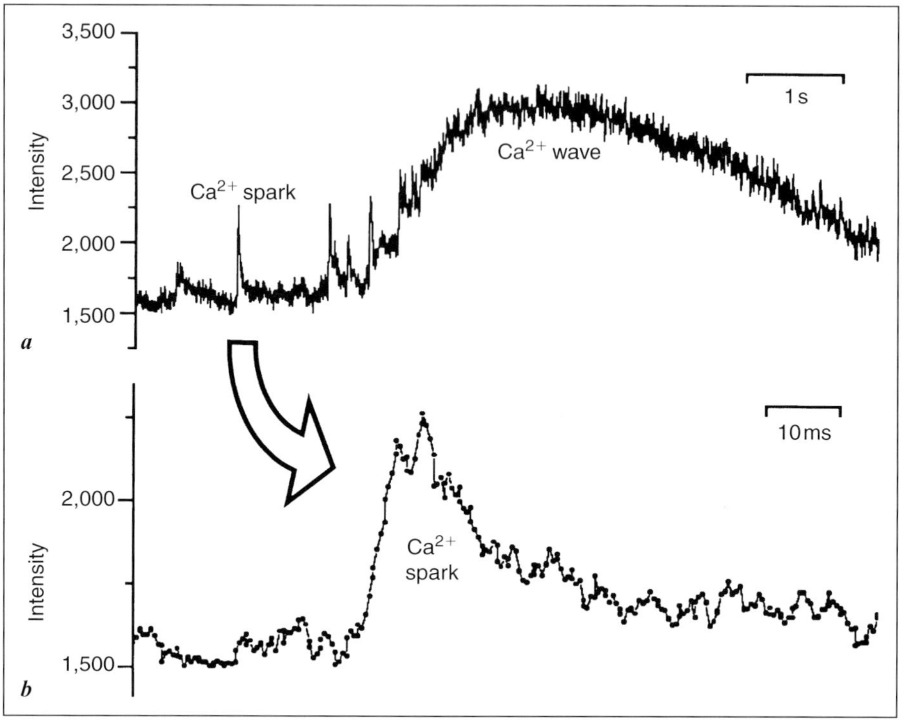

Fig. 1. Ca^{2+} sparks and a Ca^{2+} wave in smooth muscle cells of the spiral modiolar artery. Intensity expressed in arbitrary units. Confocal measurements of the cytosolic Ca^{2+} concentration with the fluorescent indicator dye fluo-4. An increase in the intensity indicates an increase in the cytosolic Ca^{2+} concentration. ***a*** Contracted time scale. Note the occurrence of several Ca^{2+} sparks and of a Ca^{2+} wave that takes about 6 s. ***b*** Expanded time scale. Note that the Ca^{2+} spark takes no more than 40–50 ms.

Ca^{2+} channel, which control the global Ca^{2+} concentration and thereby the level of myofilament constriction [22]. Several pharmacologic experiments were necessary to demonstrate that such a vasodilatory mechanism is present in the spiral modiolar artery (fig. 2). First, low concentrations of nifedipine known to selectively inhibit L-type voltage-gated Ca^{2+} channels have been shown to cause a vasodilation of the spiral modiolar artery [3]. This finding demonstrates that L-type voltage-gated Ca^{2+} channels set the vascular tone of the vessel and imply that inhibition of these channels may be part of a physiological mechanism leading to a vasodilation. Second, low concentrations of ryanodine known to stimulate Ca^{2+} release from ryanodine receptors have been shown to cause a vasodilation of the spiral modiolar artery [20, 21]. Third, low concentrations of iberiotoxin known to selectively inhibit Ca^{2+}-activated K^+ channels have

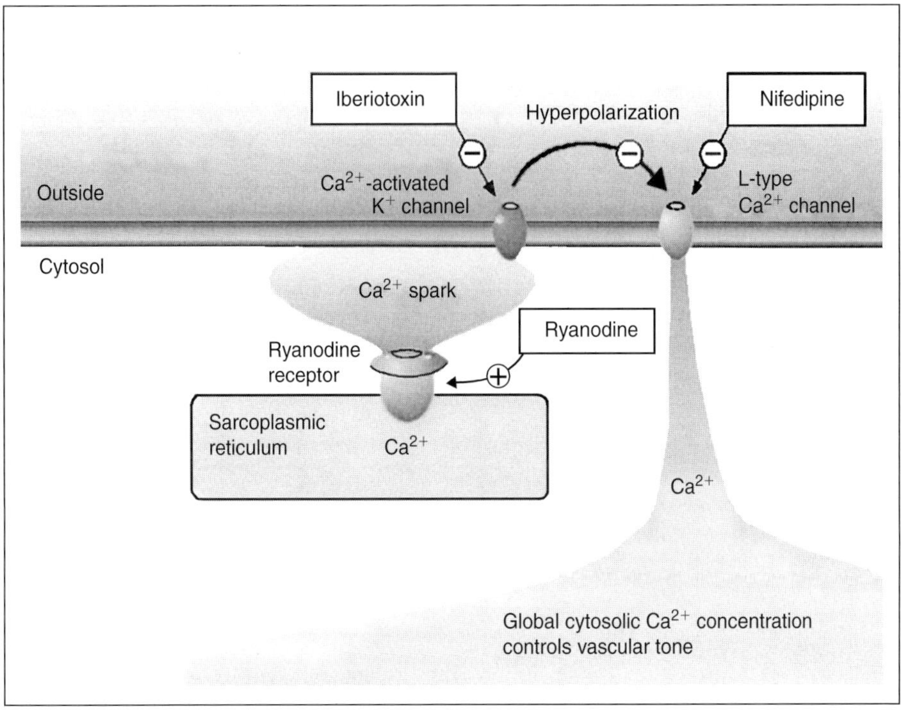

Fig. 2. Model of the signaling cascade involving the ryanodine receptor, Ca^{2+}-activated K^+ channels and L-type voltage-gated Ca^{2+} channels. Activation of the ryanodine receptor leads to Ca^{2+} sparks, opening of the Ca^{2+}-activated K^+ channels, a hyperpolarization of the membrane potential and voltage inhibition of the L-type voltage-gated Ca^{2+} channels. The inhibition of the L-type Ca^{2+} channels causes a reduction of the global cytosolic Ca^{2+} concentration and thereby a reduction of the vascular tone or vasodilation.

been shown to have no significant effect on vascular tone in the absence of ryanodine but to cause a significant vasoconstriction in the presence of ryanodine [21]. These findings demonstrate that Ca^{2+}-activated K^+ channels do not contribute to the vascular tone at resting conditions but are involved in mediating the ryanodine-induced vasodilation. Taken together, these data suggest that the spiral modiolar artery contains a mechanism, which via stimulation of ryanodine receptors, Ca^{2+} sparks, opening of Ca^{2+}-activated K^+ channels and inhibition of L-type voltage-gated Ca^{2+} channels leads to a reduction of the global cytosolic Ca^{2+} concentration and a vasodilation. It remains to be determined under which conditions this cytosolic signaling cascade is activated and which neurotransmitter, endocrine or paracrine signal may be linked to this cascade.

In summary, ongoing investigations of the isolated spiral modiolar artery are aimed to provide a detailed understanding of mechanisms contributing to cochlear blood flow regulation. These investigations may reveal new drug targets that can be employed to improve cochlear blood flow and tissue oxygenation and that provide an aid in the preservation of hearing.

Acknowledgements

The support by Research Grant R01-DC-04280 from the National Institute on Deafness and Other Communication Disorders, National Institutes of Health, is gratefully acknowledged.

References

1 Wangemann P, Gruber DD: The isolated in-vitro perfused spiral modiolar artery: Pressure dependence of vasoconstriction. Hear Res 1998;115:113–118.
2 Lamm K, Zajic G, Schacht J: Living isolated cells from inner ear vessels: A new approach for studying the regulation of cochlear microcirculation and vascular permeability. Hear Res 1994; 81:83–90.
3 Wangemann P, Cohn ES, Gruber DD, Gratton MA: Ca^{2+}-dependence and nifedipine-sensitivity of vascular tone and contractility in the isolated superfused spiral modiolar artery in vitro. Hear Res 1998;118:90–100.
4 Carlisle L, Aberdeen J, Forge A, Burnstock G: Neural basis for regulation of cochlear blood flow: Peptidergic and adrenergic innervation of the spiral modiolar artery of the guinea pig. Hear Res 1990;43:107–113.
5 Pillsbury HC, Pulver S, Carrasco VN, Scruggs K, Carver D, de Serres L, et al: Glyoxylic acid in the study of autonomic innervation in the gerbil cochlea. Arch Otolaryngol Head Neck Surg 1992; 118:413–416.
6 Brechtelsbauer PB, Prazma J, Garrett CG, Carrasco VN, Pillsbury HC: Student Research Award 1990: Catecholaminergic innervation of the inner ear. Otolaryngol Head Neck Surg 1990;103: 566–574.
7 Laurikainen EA, Ren T, Miller JM, Nuttall AL, Quirk WS: The tonic sympathetic input to the cochlear vasculature in guinea pig. Hear Res 1997;105:141–145.
8 Laurikainen EA, Kim D, Didier A, Ren T, Miller JM, Quirk WS, et al: Stellate ganglion drives sympathetic regulation of cochlear blood flow. Hear Res 1993;64:199–204.
9 Ren T-Y, Laurikainen EA, Quirk WS, Miller JM, Nuttall AL: Effects of electrical stimulation of the superior cervical ganglion on cochlear blood flow in guinea pig. Acta Otolaryngol (Stockh) 1993; 113:146–151.
10 Hultcrantz E, Angelborg C: Cochlear blood circulation studied with microspheres. Otorhinolaryngology 1978;40:65–76.
11 Ohlsen KA, Baldwin DL, Nuttall AL, Miller JM: Influence of topically applied adrenergic agents on cochlear blood flow. Circ Res 1991;69:509–518.
12 Liu S-Y, Pitovski DZ, Shivapuja BG: $Alpha_1$-adrenergic receptors in the mammalian cochlea. Acta Otolaryngol (Stockh) 1996;116:710–713.
13 Carrasco VN, Prazma J, Faber JE, Triana RJ, Pillsbury HC: Cochlear microcirculation: Effect of adrenergic agonists on arteriole diameter. Arch Otolaryngol Head Neck Surg 1990;116: 411–417.
14 Laurikainen EA, Costa O, Miller JM, Nuttall AL, Ren TY, Masta R, et al: Neuronal regulation of cochlear blood flow in the guinea-pig. J Physiol (Lond) 1994;480:563–573.

15 Wonneberger K, Wangemann P: TTX-sensitive neurogenic vasoconstriction is lacking from in vitro studied basilar artery, AICA and spiral modiolar artery of gerbil and guinea pig. Assoc Res Otolaryngol 2000;23:696–696.
16 Jiang ZG, Qiu J, Ren T, Nuttall AL: Membrane properties and the excitatory junction potentials in smooth muscle cells of cochlear spiral modiolar artery in guinea pigs. Hear Res 1999;138: 171–180.
17 Gruber DD, Dang H, Shimozono M, Scofield MA, Wangemann P: $Alpha_{1A}$ adrenergic receptors mediate vasoconstriction of the isolated spiral modiolar artery in vitro. Hear Res 1998;119: 113–124.
18 Frame MD, Sarelius IH: Arteriolar bifurcation angles vary with position and when flow is changed. Microvasc Res 1993;46:190–205.
19 Gruber DD, Dang H, Shimozono M, Scofield MA, Wangemann P. $Alpha_{1A}$ adrenergic receptors mediate vasoconstriction of the isolated spiral modiolar artery in vitro. Assoc Res Otolaryngol 1998;21:112.
20 Wonneberger K, Scofield MA, Wangemann P: Evidence for a calcium sensing receptor in the vascular smooth muscle cells of the spiral modiolar artery. J Membr Biol 2000;175:203–212.
21 Wangemann P, Berge S, Scherer EQ, Herzog M: Ca^{2+} sparks mediate vasodilation of the spiral modiolar artery via iberiotoxin-sensitive K^+ channels. Assoc Res Otolaryngol 2001;24–29.
22 Jaggar JH, Porter VA, Lederer WJ, Nelson MT: Calcium sparks in smooth muscle. Am J Physiol Cell Physiol 2000;278:C235–C256.

Dr. Philine Wangemann, Associate Professor,
Anatomy and Physiology Department, College of Veterinary Medicine,
Kansas State University, 1600 Dension Avenue, Manhattan, KS 66506 (USA)
Tel. +1 785 532 4863, Fax +1 785 532 4557, E-Mail wange@vet.ksu.edu

Felix D, Oestreicher E (eds): Rational Pharmacotherapy of the Inner Ear.
Adv Otorhinolaryngol. Basel, Karger, 2002, vol 59, pp 58–65

ET_A Receptors in the Gerbil Spiral Modiolar Artery

Elias Q. Scherer, Philine Wangemann

Anatomy and Physiology Department, Kansas State University, Manhattan, Kans., USA

Abstract

A reduction of blood flow in the spiral modiolar artery (SMA), which supplies the cochlea, is implicated in hearing loss and tinnitus. Endothelins are known to be the most potent endogenous vasoconstrictors. The purpose of the present study was to determine whether the SMA responds to endothelin, which receptor type is present and which signal transduction pathway is involved. The SMA was isolated from the gerbil cochlea by microdissection and superfused in a bath chamber on the stage of an inverted microscope. The vascular diameter was measured by video microscopy, and the cytosolic Ca^{2+} concentration was monitored simultaneously by fluo-4 fluorescence microscopy. ET-1 and ET-3 caused a dose-dependent vasoconstriction with ET-1 being the more potent agonist. The agonist sarafotoxin S6c had no significant effect. The preferential ET_A receptor antagonist BQ123 had a higher affinity inhibiting the ET-1-induced vasoconstriction than the preferential ET_B receptor antagonist BQ788. The ET-1-induced vasoconstriction was prevented by inhibition of phospholipase C with U73122. Blockade of the inositol 1,4,5-trisphosphate (IP_3) receptor on the sarcoplasmic reticulum Ca^{2+} stores with 2-aminoethoxydiphenyl borate and depletion of Ca^{2+} stores by inhibition of the sarcoplasmic Ca^{2+}-ATPase with thapsigargin prevented ET-1-induced cytosolic Ca^{2+} increase and reduced the ET-1-induced vasoconstriction. These results demonstrate that endothelin causes a vasoconstriction of the SMA, which is mediated via ET_A receptors. The data suggest that the signal transduction pathway of the ET_A receptor involves phospholipase C, IP_3 receptors and release of Ca^{2+} from thapsigargin-sensitive Ca^{2+} stores.

The spiral modiolar artery (SMA) originates via the anterior inferior cerebellar artery from the basilar artery and provides the main blood supply to the cochlea. This vessel is of great interest since alterations of blood flow along this arteriole are thought to be involved in the pathogenesis of hearing loss and tinnitus. In general, the vascular diameter controls blood flow, which, according to the law of Hagen-Poiseuille, is related to the fourth power of the vessel diameter. The vascular diameter depends on the contractile state of the vascular smooth muscle cells. More than a decade after their first description [1, 2], endothelins are still among the most potent mammalian vasoconstrictors. Endothelins are oligopeptides that occur in three isoforms, ET-1, ET-2 and ET-3. ET-1, the most abundant isoform in the vasculature, is secreted mainly by endothelial cells and in small amounts by smooth muscle cells. ET-1 is not stored in secretory granules within endothelial cells, and important stimuli such as hypoxia, ischemia and shear stress induce the transcription of ET-1 messenger RNA and the synthesis and secretion of ET-1 within minutes [3]. Endothelins are released in a paracrine/autocrine fashion from the endothelial cells. As much as 75% of ET-1 secretion from endothelial cells is directed toward the abluminal space [4], where ET-1 interacts with the endothelin receptors on smooth muscle cells, pericytes, perivascular nerves and on the endothelial cells themselves [5, 6]. The paracrine secretion of endothelins allows the vessel to control the vascular tone and thus blood flow. There are two endothelin receptor subtypes, which mediate vasoconstriction: ET_A and ET_B receptors. ET_A receptors, which are most abundant on vascular smooth muscle cells, are characterized by ET-1 being 10-fold more potent than ET-3 [7]. ET_B receptors are expressed in the vasculature predominantly by endothelial cells [8] where they induce NO synthesis and cause vasodilation [9–11]. Some vessels express ET_B receptors in the smooth muscle cells where they mediate a vasoconstriction [12–14]. ET_B receptors are characterized by an equal potency of ET-1 and ET-3.

Endothelins have been shown to cause a vasoconstriction and a reduction in blood flow in a variety of organs including the cochlea [15–18]. In particular, endothelins have been shown to cause a vasoconstriction of the capillaries in the spiral ligament, which is mediated via ET_A receptors [19]. Endothelins are also present in the vascular wall of the SMA [20]. Preliminary evidence obtained in smooth muscle cells cultured from the SMA points toward the presence of endothelin receptors and a signal transduction pathway involving Ca^{2+} [21]. The purpose of the present study was to determine whether the SMA responds to endothelin, which receptor type is present and which signal transduction pathway is involved.

ET-3- and ET-3-Induced Vasoconstriction of the SMA

The SMA was isolated from the gerbil cochlea by microdissection and superfused in a bath chamber on the stage of an inverted microscope. The vascular diameter was measured by video microscopy. The cytosolic Ca^{2+} concentration

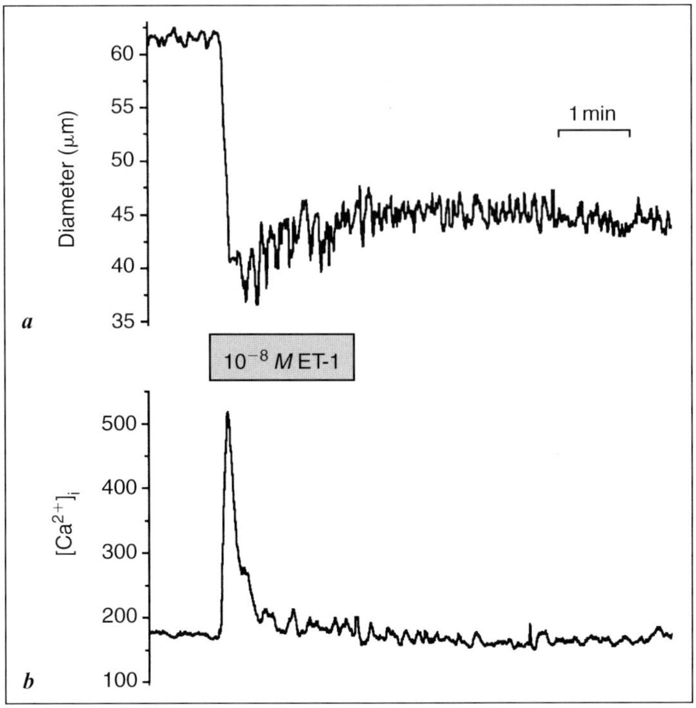

Fig. 1. Effect of ET-1 on the vascular diameter (***a***) and the cytosolic Ca^{2+} concentration (***b***) of the SMA. The vascular diameter was measured by video microscopy and the cytosolic Ca^{2+} concentration was estimated as fluo-4 fluorescence. Note that ET-1 caused a strong and lasting vasoconstriction and a fast but transient increase in the cytosolic Ca^{2+} concentration.

was monitored simultaneously by fluo-4 fluorescence microscopy. ET-1 induced a strong and long-lasting vasoconstriction of the gerbil SMA (fig. 1). The initial constriction was paralleled by a transient, fast cytosolic Ca^{2+} increase. The cytosolic Ca^{2+} concentration returned to almost resting levels, while the constriction was maintained.

Evidence for the ET_A Receptor

A comparison of the potency of ET-1 and ET-3 provides a means to distinguish between ET_A and ET_B receptors. The EC_{50} of the ET-1-induced vasoconstriction was a decade lower than that of ET-3. The higher potency of ET-1 is consistent with the presence of ET_A receptors [22]. It is also possible to distinguish between ET_A and ET_B receptors using selective ET_A and ET_B receptor antagonists.

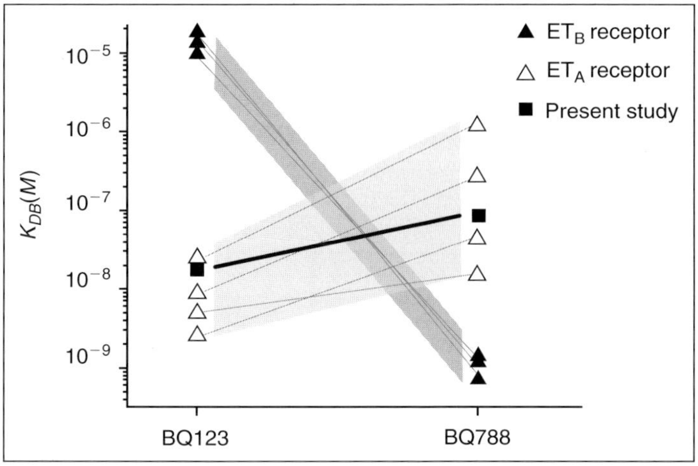

Fig. 2. Comparison of affinity values (K_{DB}) for the endothelin antagonists BQ123 and BQ788. Affinity constants obtained in the SMA are plotted (■) together with affinity constants from tissues or cells in which the endothelin receptor is known (▲, △). Note that a receptor-specific pattern emerges for the ET_A and ET_B receptors. Further, note that the K_{DB} values from the present study match the pattern for ET_A receptors.

The ET_A receptor antagonist used in this study was BQ123. BQ123 is the first and best-characterized ET_A receptor antagonist [23–25]. BQ788 was chosen as the selective ET_B receptor antagonist [26]. Both of these drugs have been used extensively to characterize ET_A and ET_B receptors in expression systems and in native tissues. The ET-1-induced vasoconstriction was competitively inhibited by the preferential ET_A receptor antagonist BQ123. The affinity constant for BQ123 was 24 nM. The preferential ET_B receptor antagonist BQ788 also competitively inhibited the ET-1-induced vasoconstriction. The affinity constant for BQ788 was 77 nM. Figure 2 shows the K_{DB} values for BQ123 and BQ788 and compares them to K_{DB} values obtained from tissues known to express ET_A or ET_B receptors [26, 27]. A receptor-specific pattern emerges from the data obtained in tissues or cells known to express ET_A or ET_B receptors. The K_{DB} values obtained in the SMA clearly follow the pattern of ET_A receptors. This observation demonstrates that the ET-1-induced vasoconstriction in the SMA is mediated by ET_A receptors. The possibility of a small subpopulation of ET_B receptors participating in the ET-1-induced vasoconstriction was excluded using sarafotoxin S6c, a selective ET_B receptor agonist [5, 28–30]. Increasing concentrations of S6c up to 10^{-7} M were without a significant effect. This finding makes it unlikely that the SMA contains a physiologically relevant subpopulation of ET_B receptors.

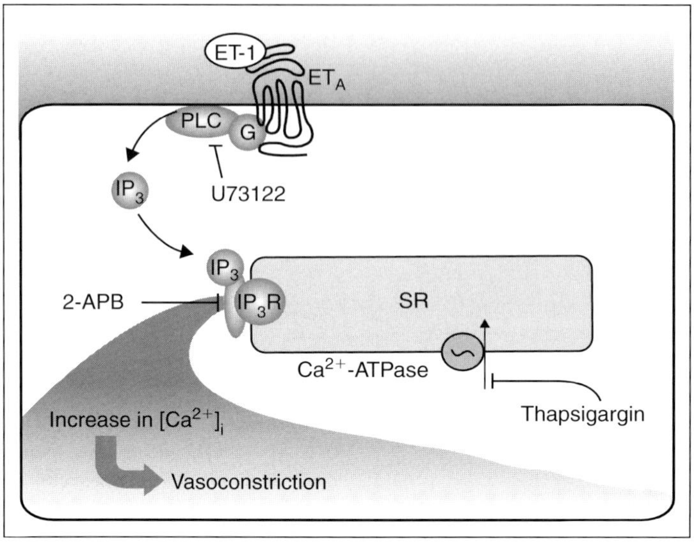

Fig. 3. Schema of the signal transduction pathway for the ET_A receptor in vascular smooth muscle cells: ET-1 binds to the ET_A receptor (ET_A), which activates $G_{q/11}$ (G) and phospholipase C (PLC) leading to the generation of IP_3 and diacylglycerol (not shown). IP_3 binds to the IP_3 receptor Ca^{2+} channel (IP_3R) permitting a fast Ca^{2+} release into the cytosol, which initializes the vasoconstriction. This signal transduction pathway was probed with several pharmacological tools: U73122 inhibits phospholipase C, 2-APB inhibits the IP_3 receptor and thapsigargin inhibits the Ca^{2+}-ATPase, which refills the sarcoplasmic reticulum (SR) Ca^{2+} store (*see text*).

Signal Transduction Pathway of ET_A Receptors in the SMA

The proposed signal transduction pathway for ET_A receptors in vascular smooth muscle cells is drawn schematically in figure 3. It has been shown that ET_A receptors signal via $G_{q/11}$ and phospholipase C [31], leading to a generation of inositol 1,4,5-trisphosphate (IP_3) and diacylglycerol [32]. IP_3 binds to the IP_3 receptor on the sarcoplasmic reticulum. The IP_3 receptor is coupled to a Ca^{2+} channel. Binding of IP_3 opens this Ca^{2+} channel permitting a fast Ca^{2+} release into the cytosol, which initializes the constriction of the smooth muscle cell. The high Ca^{2+} concentration in the sarcoplasmic reticulum, which is the main cytosolic Ca^{2+} store, is maintained by a Ca^{2+}-ATPase, which transports Ca^{2+} from the cytosol into the sarcoplasmic reticulum. When this Ca^{2+}-ATPase is inhibited with the plant alkaloid thapsigargin, the sarcoplasmic reticulum looses its high Ca^{2+} concentration. The consequence of the Ca^{2+} depletion of the sarcoplasmic reticulum is that the smooth muscle cell looses its ability to respond to IP_3 with a cytosolic Ca^{2+} increase.

This known signaling cascade was experimentally verified in the SMA. Phospholipase C was inhibited selectively with 1 μ*M* U73122 [33, 34]. Preincubation of the vessels with U73122 for 10 min prevented the ET-1-induced constriction. Applying U73122 after constricting the vessels with ET-1 released the constriction within 6 min. The specificity of U73122 was confirmed by a control experiment using 1 μ*M* U73343. U73343 had no significant effect. This finding strengthens the assumption that U73122 inhibited phospholipase C and strengthens the conclusion that phospholipase C plays a significant role in the signaling cascade of the ET_A receptor.

The involvement of the IP_3 receptor in the signaling cascade was verified using a selective blocker of the IP_3 receptor, 75 μ*M* 2-aminoethoxydiphenyl borate (2-APB) [35]. Preincubation of the vessels for 90 s with 75 μ*M* 2-APB prevented ET-1-induced Ca^{2+} release from cytosolic Ca^{2+} stores and reduced the ET-1-induced constriction.

Last but not least, the involvement of the cytosolic Ca^{2+} stores in the ET-1-induced constriction was evaluated. The sarcoplasmic reticulum can be depleted of its Ca^{2+} by inhibiting the sarcoplasmic reticulum Ca^{2+}-ATPase with 1 μ*M* thapsigargin. Preincubation of the vessels with 1 μ*M* thapsigargin prevented the ET-1-induced Ca^{2+} release from cytosolic Ca^{2+} stores and reduced the ET-1-induced constriction. This observation confirms the involvement of the Ca^{2+} store.

In conclusion, endothelins cause a strong and long-lasting vasoconstriction of the SMA, which is mediated via ET_A receptors. The signaling cascade for these ET_A receptors involves activation of $G_{q/11}$ and phospholipase C, release of IP_3 and mobilization of Ca^{2+} from cytosolic, thapsigargin-sensitive Ca^{2+} stores. The observation that the ET-1-induced vasoconstriction was very strong and long-lasting suggests that ET-1 may play a major role in the pathogenesis of syndromes such as sudden hearing loss and tinnitus.

Acknowledgments

The support by Research Grant RO1-DC04280 from the National Institute on Deafness and Other Communication Disorders, National Institutes of Health, is gratefully acknowledged.

References

1 Inoue A, Yanagisawa M, Kimura S, Kasuya Y, Miyauchi T, Goto K, et al: The human endothelin family: Three structurally and pharmacologically distinct isopeptides predicted by three separate genes. Proc Natl Acad Sci USA 1989;86:2863–2867.

2 Yanagisawa M, Kurihara H, Kimura S, Tomobe Y, Kobayashi M, Mitsui Y, et al: A novel potent vasoconstrictor peptide produced by vascular endothelial cells. Nature 1988;332:411–415.

3 Nakamura S, Naruse M, Naruse K, Demura H, Uemura H: Immunocytochemical localization of endothelin in cultured bovine endothelial cells. Histochemistry 1990;94:475–477.
4 Yoshimoto S, Ishizaki Y, Sasaki T, Murota S: Effect of carbon dioxide and oxygen on endothelin production by cultured porcine cerebral endothelial cells. Stroke 1991;22:378–383.
5 Rubanyi GM, Polokoff MA: Endothelins: Molecular biology, biochemistry, pharmacology, physiology, and pathophysiology. Pharmacol Rev 1994;46:325–415.
6 Pollock DM, Keith TL, Highsmith RF: Endothelin receptors and calcium signaling. FASEB J 1995;9:1196–1204.
7 Arai H, Hori S, Aramori I, Ohkubo H, Nakanishi S: Cloning and expression of a cDNA encoding an endothelin receptor. Nature 1990;348:730–732.
8 Sakurai T, Yanagisawa M, Takuwa Y, Miyazaki H, Kimura S, Goto K, et al: Cloning of a cDNA encoding a non-isopeptide-selective subtype of the endothelin receptor. Nature 1990;348:732–735.
9 Hirata Y, Emori T, Eguchi S, Kanno K, Imai T, Ohta K, et al: Endothelin receptor subtype B mediates synthesis of nitric oxide by cultured bovine endothelial cells. J Clin Invest 1993;91:1367–1373.
10 Fujitani Y, Ueda H, Okada T, Urade Y, Karaki H: A selective agonist of endothelin type B receptor, IRL 1620, stimulates cyclic GMP increase via nitric oxide formation in rat aorta. J Pharmacol Exp Ther 1993;267:683–689.
11 Tsukahara H, Ende H, Magazine HI, Bahou WF, Goligorsky MS: Molecular and functional characterization of the non-isopeptide-selective ETB receptor in endothelial cells: Receptor coupling to nitric oxide synthase. J Biol Chem 1994;269:21778–21785.
12 Pang CY, Zhang J, Xu H, Lipa JE, Forrest CR, Neligan PC: Role and mechanism of endothelin-B receptors in mediating ET-1-induced vasoconstriction in pig skin. Am J Physiol 1998;275:R1066–R1074.
13 LaDouceur DM, Flynn MA, Keiser JA, Reynolds E, Haleen SJ: ETA and ETB receptors coexist on rabbit pulmonary artery vascular smooth muscle mediating contraction. Biochem Biophys Res Commun 1993;196:209–215.
14 Haynes WG, Strachan FE, Webb DJ: Endothelin ETA and ETB receptors cause vasoconstriction of human resistance and capacitance vessels in vivo. Circulation 1995;92:357–363.
15 Luscher TF: Endothelin, endothelin receptors, and endothelin antagonists. Curr Opin Nephrol Hypertens 1994;3:92–98.
16 Barton M, Luscher TF: Endothelin antagonists for hypertension and renal disease. Curr Opin Nephrol Hypertens 1999;8:549–556.
17 Oemar BS, Yang Z, Luscher TF: Molecular and cellular mechanisms of atherosclerosis. Curr Opin Nephrol Hypertens 1995;4:82–91.
18 Quirk WS, Laurikainen EA, Avinash G, Nuttall AL, Miller JM: The role of endothelin on the regulation of cochlear blood flow. Assoc Res Otolaryngol 1992;15:37.
19 Sadanaga M, Liu J, Wangemann P: Endothelin-A receptors mediate vasoconstriction of capillaries in the spiral ligament. Hear Res 1997;112:106–114.
20 Jinnouchi K, Tomiyama S, Pawankar R, Ikezono T, Yagi T: Distribution of endothelin-1-like activity in the cochlea of normal guinea pigs. Acta Otolaryngol (Stockh) 1997;117:41–45.
21 Lamm K, Zajic G, Arnold W, Schacht J: Identification of vasoactive receptors in the vital isolated cochlear spiral modiolar artery. Assoc Res Otolaryngol 1995;18:568.
22 Scherer EQ, Wonneberger K, Wangemann P: Endothelin-A receptors mediate vasoconstriction of the gerbil spiral modiolar artery. Assoc Res Otolaryngol 2000;23:5817.
23 Ihara M, Yamanaka R, Ohwaki K, Ozaki S, Fukami T, Ishikawa K, et al: [^{3}H]BQ-123, a highly specific and reversible radioligand for the endothelin ETA receptor subtype. Eur J Pharmacol 1995;274:1–6.
24 Ihara M, Noguchi K, Saeki T, Fukuroda T, Tsuchida S, Kimura S, et al: Biological profiles of highly potent novel endothelin antagonists selective for the ETA receptor. Life Sci 1992;50:247–255.
25 Ihara M, Ishikawa K, Fukuroda T, Saeki T, Funabashi K, Fukami T, et al: In vitro biological profile of a highly potent novel endothelin (ET) antagonist BQ-123 selective for the ETA receptor. J Cardiovasc Pharmacol 1992;20(suppl 12):S11–S14.

26 Ishikawa K, Ihara M, Noguchi K, Mase T, Mino N, Saeki T, et al: Biochemical and pharmacological profile of a potent and selective endothelin B-receptor antagonist, BQ-788. Proc Natl Acad Sci USA 1994;91:4892–4896.
27 Ozaki S, Ihara M, Saeki T, Fukami T, Ishikawa K, Yano M: Endothelin ETB receptors couple to two distinct signaling pathways in porcine kidney epithelial LLC-PK1 cells. J Pharmacol Exp Ther 1994;270:1035–1040.
28 Kloog Y, Ambar I, Sokolovsky M, Kochva E, Wollberg Z, Bdolah A: Sarafotoxin, a novel vasoconstrictor peptide: Phosphoinositide hydrolysis in rat heart and brain. Science 1988;242: 268–270.
29 Williams DL Jr, Jones KL, Colton CD, Nutt RF: Identification of high affinity endothelin-1 receptor subtypes in human tissues. Biochem Biophys Res Commun 1991;180:475–480.
30 Williams DL Jr, Jones KL, Pettibone DJ, Lis EV, Clineschmidt BV: Sarafotoxin S6c: An agonist which distinguishes between endothelin receptor subtypes. Biochem Biophys Res Commun 1991;175:556–561.
31 Pang DC, Johns A, Patterson K, Botelho LH, Rubanyi GM: Endothelin-1 stimulates phosphatidylinositol hydrolysis and calcium uptake in isolated canine coronary arteries. J Cardiovasc Pharmacol 1989;13(suppl 5):S75–S79.
32 Rapoport RM, Stauderman KA, Highsmith RF: Effects of EDCF and endothelin on phosphatidylinositol hydrolysis and contraction in rat aorta. Am J Physiol 1990;258:C122–C131.
33 Yule DI, Williams JA: U73122 inhibits Ca^{2+} oscillations in response to cholecystokinin and carbachol but not to JMV-180 in rat pancreatic acinar cells. J Biol Chem 1992;267:13830–13835.
34 Smith RJ, Sam LM, Justen JM, Bundy GL, Bala GA, Bleasdale JE: Receptor-coupled signal transduction in human polymorphonuclear neutrophils: Effects of a novel inhibitor of phospholipase C-dependent processes on cell responsiveness. J Pharmacol Exp Ther 1990;253:688–697.
35 Ma HT, Patterson RL, van Rossum DB, Birnbaumer L, Mikoshiba K, Gill DL: Requirement of the inositol trisphosphate receptor for activation of store-operated Ca^{2+} channels. Science 2000;287: 1647–1651.

Dr. Philine Wangemann, Associate Professor, Anatomy and Physiology Department,
Kansas State University, 1600 Dension Avenue, Manhattan, KS 66506 (USA)
Tel. +1 785 532 4863, Fax +1 785 532 4557, E-Mail Wange@vet.ksu.edu

Felix D, Oestreicher E (eds): Rational Pharmacotherapy of the Inner Ear.
Adv Otorhinolaryngol. Basel, Karger, 2002, vol 59, pp 66–74

Immunological Damage to the Inner Ear: Current and Future Therapeutic Strategies

Allen F. Ryan[a,b], *Kwang Pak*[a], *Wesley Low*[a], *Alex Battaglia*[a], *Lina Mullen*[a], *Jeffrey P. Harris*[a], *Elizabeth M. Keithley*[a]

Departments of [a] Surgery/Otolaryngology and [b] Neurosciences,
University of California, San Diego, and VA Medical Center, La Jolla, Calif., USA

Abstract

There is considerable evidence to suggest that hearing and vestibular function can be influenced by immunity in the inner ear. While immunity can protect against infections of the labyrinth, immune response also has the capacity to damage the delicate tissues of the inner ear. Antigenic challenge of the inner ear of sensitized animals leads to rapid accumulation of leukocytes, antibody production, hearing loss and tissue damage. Moreover, a number of systemic autoimmune disorders include hearing loss and vertigo as part of their constellation of symptoms. It also appears that autoimmune damage can exist as an entity confined to the labyrinth. Immune disorders of the inner ear are of special interest since they are among the few forms of hearing loss that are currently amenable to medical treatment. In addition, recent developments in understanding the intracellular pathways that participate in damage to the inner ear provide new opportunities for pharmacotherapy of immune-mediated disorders of hearing and balance.

Labyrinthine Immunity and Immune-Mediated Damage

Traditionally, the inner ear was viewed as an immunologically privileged site, similar to the brain, in that it was separated from cellular and humoral immunity by a blood-labyrinthine barrier. In this view, the inner ear would be subject to immunity only during catastrophic breakdown of its integrity, as in bacterial or viral labyrinthitis. However, experiments in the 1980s demonstrated that the inner ear can participate quite vigorously in immune response, even in the absence of infection.

Harris [1] and Harris et al. [2] found that antigen introduced into the cochlear perilymph of naive guinea pigs produces a brisk systemic immune response, which is in fact more vigorous than that produced by immunization of the middle ear, indicating that the afferent limb of systemic immunity functions well in the labyrinth. Moreover, antigen introduced into the cochlea of systemically immunized animals can result in significant sensorineural hearing loss (SNHL), due to a vigorous secondary immune response [3, 4]. This response includes extravasation of leukocytes including both T and B lymphocytes, via the spiral modiolar vein [5]. Inflammatory tissue damage and local antibody production also occur. Inner ear immune responses are strongly linked to the endolymphatic sac, since its ablation, or even blockage of the endolymphatic duct, reduces inner ear immune responses [6]. Immunosuppression of naive animals can reduce inflammation and SNHL due to viral labyrinthitis, suggesting that bystander injury to cochlear cells can be caused by the primary host response to virus [7]. Of course, as in other systems, immunity in the inner ear is a double-edged sword. While it can damage the labyrinth, immunity can also serve a protective function at this site, as it does elsewhere. For example, it has been demonstrated that preexisting systemic immunity can protect the inner ear from viral infection [8].

In 1979, McCabe [9] introduced the concept that autoimmunity might damage the labyrinth. He presented a series of patients with bilateral, progressive SNHL that improved following corticosteroid treatment [9]. Autoimmunity has since been proposed as an etiology for Ménière's disease [10], sudden SNHL [11] and acute vertigo [12]. Several systemic autoimmune disorders, including polyarteritis nodosa, systemic lupus erythematosus (SLE), relapsing polychodritis, ulcerative colitis and Wegener's granulomatosis, also produce auditory and vestibular symptoms, perhaps mediated by vasculitis [13]. Some patients with suspected autoimmune inner ear disease (AIED) have later developed systemic autoimmune disease [14]. Recently, Sone et al. [15] have assessed 14 temporal bones from 7 individuals with SLE. The most consistent findings were hair cell and spiral ganglion cell loss. However, unusual accretions were observed in the stria vascularis of 6/14 temporal bones.

Definitive proof that SNHL and vertigo can be caused by autoimmunity to inner ear antigens has been elusive. The disorder is uncommon and the inner ear is a small, delicate organ deeply embedded in bone. Standard clinical immunological assays cannot be applied locally, tissue biopsies are unavailable and histopathological data are scant. The little histopathology published from suspected AIED patients shows labyrinthine fibrosis and/or bone deposition, consistent with late sequelae of inflammation [16]. Immune reactivity against inner ear antigens is frequently detected in suspect AIED patients, but the results are far from consistent.

A number of animal studies support an autoimmune etiology for some forms of SNHL. Animal models have been valuable adjuncts in the study of AIED, since antigen and immunization history can be rigorously controlled, and histopathology is available. Initial studies on guinea pigs immunized with bovine inner ear extracts showed hearing loss and mild inner ear inflammation in a subset of animals [17]. Recent work confirmed these findings, and indicated that hearing losses tend to be modest. Bouman et al. [18] immunized guinea pigs with swine inner ear extracts. They reported modest declines in compound action potentials, but no changes in the cochlear microphonic, 2 and 6 weeks later. This suggests that hearing loss occurred at the inner hair cell and/or spiral ganglion neuron, rather than at the outer hair cell. Hearing loss was associated with increased Western blot reactivity to 68-kD and other antigens [18]. A monoclonal antibody that specifically reacts with supporting cells in the organ of Corti has been shown to produce high-frequency hearing loss in mice carrying the hybridoma [19].

Animal models have also been used to study the relationship between systemic autoimmune disease and the inner ear. The MRL-Fas^{lpr} mouse is used as a model of SLE, due to the accumulation of autoreactive T cells normally eliminated by Fas-mediated apoptosis. This model displays progressive hearing loss. Ruckenstein et al. [20] found that the most striking inner ear pathology in this model was in the stria vascularis, with progressive, hydropic degeneration of intermediate cells, perhaps consistent with the strial pathology observed in human SLE temporal bones as described above. In addition, Ruckenstein and Hu [21] observed the deposition of both complement-fixing and non-complement-fixing antibodies in capillaries of the stria vascularis and, to a lesser extent, in other structures. This was not associated with signs of inflammation, however. The same group found that systemic dexamethasone suppressed antibody deposition within the stria but failed to suppress strial degeneration [20]. In contrast, Wobig et al. [22] found that systemic prednisolone protected hearing in MRL-Fas^{lpr} mice, suggesting that factors other than strial degeneration contribute to SNHL in this model. The Palmerston-North mouse is also employed as a model of SLE with hearing loss. These animals develop abnormal mineralization of connective tissue in the region of the 8th nerve root within the modiolus. However, there is no deposition of antibody or cellular infiltration in the cochlea [23].

Pharmacotherapy for Immune-Mediated Inner Ear Disorders

As noted by McCabe [9] and a number of subsequent authors, immune-mediated inner ear disorders are unique in that medical therapy can reverse

SNHL. Once a diagnosis of immune-mediated hearing loss is established or considered highly presumptive in a patient, high-dose prednisone is the mainstay of treatment [24]. Early institution of 60 mg of prednisone daily for about a month is now widely used, as short-term or lower-dose long-term therapy has either been ineffective or fraught with the risk of relapse. Prednisone is then tapered slowly if a positive response to therapy is obtained. If during the tapering hearing suddenly falls, reinstitution of high-dose prednisone is indicated. One sensitive predictor of imminent relapse can be the appearance of loud tinnitus in one or both ears. If patients show steroid responsiveness but attempts at tapering result in relapse, the addition of a cytotoxic drug should be considered. The most widely used of these agents are methotrexate (MTX) and cyclophosphamide. The former has the advantage of being less toxic and having fewer long-term hematopoietic risks, such as the development of neoplasia [25]. MTX should be given as an oral dose of 7.5–20 mg weekly with folic acid, and the patient closely monitored for toxicity with complete blood count, platelets, blood urea nitrogen, creatinine, liver function tests and urinalysis. The prednisone sparing effects of MTX may take 1–2 months to achieve; therefore, prednisone should be maintained until such effects are obtained. If high-dose prednisone has not been effective in restoring hearing, it is unlikely that MTX will offer additional efficacy. At the time of writing, a multi-institutional clinical trial is under way to compare the efficacy of MTX and prednisone versus prednisone alone for AIED management. The results should help to delineate an appropriate therapy for suspected AIED.

For patients with severe hearing losses, positive Western blot or other assay, and nonresponsiveness to prednisone or MTX therapy, cyclophosphamide can be considered [14]. At oral doses of 1–2 mg/day taken each morning with liberal amounts of fluid, the risk of hemorrhagic cystitis or drug effects on the bladder are minimized. Again, appropriate monitoring of peripheral blood counts is required. The risk of permanent sterility should be outlined, and cyclophosphamide should not be administered to children. If no response to high-dose prednisone is achieved and the patient is negative on Western blot or other assays, continuing potentially toxic drugs may be futile. However, there are no hard and fast rules, and a practitioner might be justified in trying cytotoxic drugs on an empiric basis because unrelenting progressive deafness is a serious handicap for a previously normal-hearing person. Luetje [26] recommends plasmapheresis for difficult to manage patients, and this can be a useful adjunct to the above-mentioned immunosuppressive drugs.

Parnes et al. [27] noted that local corticosteroids appear to be more effective in the treatment of some autoimmune disorders, such as corneal inflammation due to Cogan's syndrome. They therefore investigated the pharmacokinetics of hydrocortisone, methylprednisone and dexamethasone in perilymph and endolymph after oral, intravenous or intratympanic administration. Dexamethasone was found to be largely excluded from the cochlea by the blood-labyrinthine barrier. While both methylprednisone and hydrocortisone reached inner ear fluids after systemic administration, they were substantially attenuated by the blood-labyrinthine barrier. Much higher levels of all three drugs were observed in cochlear fluids after intratympanic administration, although with rapid declines over a 6- to 24-hour period. Similar results were noted by Chandrasekhar et al. [28]. Parnes et al. [27] also reported that repeated intratympanic administration of corticosteroids in a small series of patients with hearing loss of diverse origins was followed by improvement in some cases, but no control group was included. In a controlled animal study, Yang et al. [29] found that no reduction in inflammation or hearing loss was achieved following local application of cyclosporin, prednisolone acetate, FK-506 or fluorouracil, in a model of immune-mediated SNHL. Local effects are not, of course, the only basis for the therapeutic efficacy of immunosupressants. By decreasing peripheral blood leukocytes, these agents reduce the population of cells that can be recruited to the inner ear to participate in immune and inflammatory damage.

Future Therapeutic Strategies

Immunosuppression will undoubtedly continue to be central to the treatment of immune-mediated labyrinthine disorders, and ongoing advances in systemic and local immunosuppressants [30, 31] will thus yield improvements in the treatment of inner ear disorders. Additional research on local delivery of immunosuppressives to the inner ear may also yield results. For example, the role of the endolymphatic sac in inner ear immune responses [6] suggests that this might be a potentially effective site for local pharmacotherapy.

In addition to immunotherapy, additional forms of pharmacotherapy may be useful in treating damage to the inner ear caused by immunological events. Immune-mediated hearing loss and vestibular dysfunction are caused by damage to the functional elements of the inner ear, primarily hair cells, neurons and the stria vascularis. During the past several years, significant progress has been made in understanding how damage to hair cells and neurons occurs. In particular, intracellular events that mediate at least some aspects of damage to hair cells have been discovered. For example, production of free radicals has been found to be an important factor [32]. Moreover, the death of hair cells appears to occur in some cases via an apoptotic process [33]. These observations have led to the demonstration that hair cell loss due to various potentially damaging stimuli can be prevented by the application of endogenous substances that interfere with the mechanism leading to cell damage or that interfere in apoptosis [34–36]. It has also been shown that hair cells can be protected from damage by the exposure to certain non-neurotrophic growth factors [37, 38], indicating that hair cells can be influenced by the activation of the same cell survival programs that operate in other tissues. While these observations have been directed at noise- and drug-induced SNHL, it seems likely that similar processes underlie at least some hair cell loss that occurs during immune-mediated SNHL.

With respect to cochlear neurons, swelling of the spiral ganglion dendrites due to intense noise and ischemia had been noted for decades, as was the fact that hair cell loss can lead to anterograde degeneration of primary auditory neurons, with little or no understanding of the underlying mechanisms. However, by the early 1990s, it became clear that neurotransmission between the hair cell and the spiral ganglion neuron was mediated by glutamate receptors [39, 40] and that dendritic damage could be ameliorated by glutamate antagonists [41]. In a separate series of significant advances, it was found that hair cells produce neurotrophins and other neuronal survival factors [42], and that absence of these factors can result in anterograde degeneration of primary sensory neurons [43, 44]. This in turn led to the demonstration that neuronal loss, secondary to hair cell loss, could be reduced by administration of neurotrophins [45]. Again, while immune-mediated damage to inner ear neurons undoubtedly occurs by inflammatory pathways, it is possible that at least some of the neuronal losses that occur in immune-mediated SNHL are caused by similar mechanisms.

Other investigators have observed a synergistic relationship between protection of cochlear hair cells and neurons. Basile et al. [46] observed protection of both hair cells and spiral ganglion neurons from ototoxicity by glutamate antagonists that would presumably directly affect only spiral ganglion neurons. Similarly, Duan et al. [47] found that the N-methyl-*D*-aspartate receptor antagonist MK801 provided modest protection of hearing from noise or the ototoxic antibiotic amikacin. Duan et al. [47] also combined MK801 with the neurotrophin NT-3 and found an even greater protection of hair cells and hearing, even though NT-3 alone has little or no protective effect on hair cells [45].

The pharmacological strategies that have been found to protect hair cells and inner ear neurons from damage have yet to be applied to immune-mediated inner ear damage. However, to the extent that they address final common pathways of cellular damage or activate common survival programs, such strategies could be effective. Of course, numerous hurdles must be overcome before the application of protective strategies that have been demonstrated in animal studies can be applied to humans. While strategies and devices for delivery of protective agents to the cochlea are under active development, the agents must be shown to be safe as well as efficacious. It is also possible that protective strategies effective for the acute insults typically used in animal experiments might not be as useful for chronic conditions that most often occur in patients.

However, vector-based [38, 48] and germ line [49] gene therapy have been shown to provide protection of both hair cells and auditory neurons in animals, and have the potential for long-term delivery of protective substances. Almost no attention has been devoted to preventing damage to the stria vascularis. However, this tissue is clearly a target of at least some forms of immune-mediated SNHL [15, 20, 21] and as a highly vascular tissue might be particularly suited to systemic pharmacotherapy.

Conclusions

Both animal experiments and patient studies suggest that immune processes can damage the inner ear. Despite uncertainty over the etiology and difficulties in diagnosis, patients with suspected immune-related inner ear disorders are frequently responsive to treatment with immunosuppressive drugs. Since there are few forms of inner ear disorders that can be treated medically, immune-related inner ear disorders represent a unique opportunity to reverse SNHL and vestibular disfunction. Moreover, recent progress in understanding the intracellular mechanisms of hair cell and neuronal damage suggest additional strategies for pharmacotherapy of immune-mediated inner ear disorders.

Acknowledgements

This work was supported by grants DC00129, DC00139 and DC04268 from the NIH/NIDCD, and by the Research Service of the VA.

References

1 Harris JP: Immunology of the inner ear: Response of the inner ear to antigen challenge. Otolaryngol Head Neck Surg 1983;91:18–32.
2 Harris JP, Woolf NK, Ryan AF: Elaboration of systemic immunity following inner ear immunization. Am J Otolaryngol 1985;6:148–152.
3 Woolf NK, Harris JP: Cochlear pathophysiology associated with inner ear immune responses. Acta Otolaryngol 1986;102:353–364.
4 Ma C, Billings P, Harris JP, Keithley EM: Characterization of an experimentally induced inner ear immune response. Laryngoscope 2000;110:451–456.
5 Stearns GS, Keithley EM, Harris JP: Development of high endothelial venule-like characteristics in the spiral modiolar vein induced by viral labyrinthitis. Laryngoscope 1993;103:890–898.
6 Tomiyama S, Harris JP: The role of the endolymphatic sac in inner ear immunity. Acta Otolaryngol 1987;103:182–188.
7 Darmstadt GL, Keithley EM, Harris JP: Effects of cyclophosphamide on the pathogenesis of cytomegalovirus-induced labyrinthitis. Ann Otol Rhinol Laryngol 1990;99:960–968.

8 Woolf N, Harris JP, Ryan AF, Butler DM, Richman DD: Hearing loss in experimental cytomegalovirus infection of the guinea pig inner ear: Prevention by systemic immunity. Ann Otol Rhinol Laryngol 1985;94:350–356.
9 McCabe B: Autoimmune sensorineural hearing loss. Ann Otol Rhinol Laryngol 1979;88: 585–589.
10 Hughes GB, Barna B, Kinner S, Calabrese L, Hamid M, Nalepa NJ: Autoimmune endolymphatic hydrops: A five-year review. Otolaryngol Head Neck Surg 1988;98:221–225.
11 Moskowitz D, Lee KJ, Smith HW: Steroid use in idiopathic sudden sensorineural hearing loss. Laryngoscope 1984;94:664–666.
12 Ariyusa L, Byl F, Sprague M, Adour K: The beneficial effects of methylprednisone in acute vestibular vertigo. Arch Otolaryngol Head Neck Surg 1990;166:700–703.
13 Harris JP, Ryan AF: Fundamental immune mechanisms of the brain and inner ear. Otolaryngol Head Neck Surg 1995;112:639–653.
14 McCabe BF: Autoimmune inner ear disease; in Bernstein I, Ogra P (eds): Immunology of the Ear. New York, Raven Press, 1987, pp 427–435.
15 Sone M, Schachern PA, Paparella MM, Morizono N: Study of systemic lupus erythematosus in temporal bones. Ann Otol Rhinol Laryngol 1999;108:338–344.
16 Keithley EM, Chen MC, Linthicum F: Clinical diagnoses associated with histologic findings of fibrotic tissue and new bone in the inner ear. Laryngoscope 1998;108:87–91.
17 Harris JP: Experimental autoimmune sensorineural hearing loss. Laryngoscope 1987;97:63–76.
18 Bouman H, Klis SF, Meeuwsen F, de Groot JC, Smoorenburg GF, Veldman JE: Experimental autoimmune inner ear disease: An electrocochleographic and histophysiologic study. Ann Otol Rhinol Laryngol 2000;109:457–466.
19 Nair TS, Raphael Y, Dolan DF, Parrett TJ, Perlman LS, Brahmbhatt VR, Wang Y, Hou X, Ganjei G, Nuttall AL, Altschuler RA, Carey TE: Monoclonal antibody induced hearing loss. Hear Res 1995; 83:101–113.
20 Ruckenstein MJ, Sarwar A, Hu L, Shami H, Marion TN: Effects of immunosuppression on the development of cochlear disease in the MRL-Fas(lpr) mouse. Laryngoscope 1999;109:626–630.
21 Ruckenstein MJ, Hu L: Antibody deposition in the stria vascularis of the MRL-Fas(lpr) mouse. Hearing Res 1999;127:137–142.
22 Wobig RJ, Kempton JB, Trune DR: Steroid-responsive cochlear dysfunction in the MRL/lpr autoimmune mouse. Otolaryngol Head Neck Surg 1999;121:344–347.
23 Khan DC, DeGagne JM, Trune DR: Abnormal cochlear connective tissue mineralization in the Palmerston-North autoimmune mouse. Hear Res 2000;142:12–22.
24 Harris JP, Moscicki RA, Hughes GB: Immunologic disorders of the inner ear; in Hughes GB, Pensak ML (eds): Clinical Otology, ed 2. New York, Thieme, pp 381–391.
25 Sismanis A, Thompson T, Willis HE: Methotrexate therapy for autoimmune hearing loss: A preliminary report. Laryngoscope 1994;104:932–934.
26 Luetje CM: Theoretical and practical implications for plasmapheresis in autoimmune inner ear disease. Laryngoscope 1989;99:1137–1146.
27 Parnes LS, Sun AH, Freeman DJ: Corticosteroid pharmacokinetics in the inner ear fluids: An animal study followed by clinical application. Laryngoscope 1999;109:1–17.
28 Chandrasekhar SS, Rubinstein RY, Kwartler JA, Gatz M, Connelly PE, Huang E, Baredes S: Dexamethasone pharmacokinetics in the inner ear: Comparison of route of administration and use of facilitating agents. Otolaryngol Head Neck Surg 2000;122:521–528.
29 Yang GS, Song HT, Keithley EM, Harris JP: Intratympanic immunosuppressives for prevention of immune-mediated sensorineural hearing loss. Am J Otol 2000;21:499–504.
30 Billaud EM: Clinical pharmacology of immunosuppressive drugs: Year 2000 – Time for alternatives. Therapie 2000;55:177–183.
31 Nevins TE: Overview of new immunosuppressive therapies. Curr Opin Pediatr 2000;12:146–150.
32 Hoffman DW, Whitworth CA, Jones KL, Rybak LP: Nutritional status, glutathione levels, and ototoxicity of loop diuretics and aminoglycoside antibiotics. Hear Res 1987;31:217–222.
33 Li L, Nevill G, Forge A: Two modes of hair cell loss from the vestibular sensory epithelia of the guinea pig inner ear. J Comp Neurol 1995;355:405–417.

34 Otto WC, Brown RD, Gage-White L, Kupetz S, Anniko M, Penny JE, Henley CM: Effects of cisplatin and thiosulfate upon auditory brainstem responses of guinea pigs. Hear Res 1988;35: 79–85.
35 Garetz SL, Altschuler RA, Schacht J: Attenuation of gentamicin ototoxicity by glutathione in the guinea pig in vivo. Hear Res 1994;77:81–87.
36 Song BB, Sha SH, Schacht J: Iron chelators protect from aminoglycoside-induced cochleo- and vestibulo-toxicity. Free Radic Biol Med 1998;25:189–195.
37 Low W, Dazert S, Baird A, Ryan AF: Basic fibroblast growth factor (FGF-2) protects rat cochlear hair cells in organotypical culture from aminoglycoside injury. J Cell Physiol 1996;167:443–450.
38 Yagi M, Magal E, Sheng Z, Ang KA, Raphael Y: Hair cell protection from aminoglycoside ototoxicity by adenovirus-mediated overexpression of glial cell line-derived neurotrophic factor. Hum Gene Ther 1999;10:813–823.
39 Ryan AF, Brumm D, Kraft M: Occurrence and distribution of non-NMDA glutamate receptor mRNAs in the cochlea. Neuroreport 1991;2:643–646.
40 Eybalin M: Neurotransmitters and neuromodulators of the mammalian cochlea. Physiol Rev 1993;73:309–373.
41 Pujol R, Puel JL, Eybalin M: Implication of non-NMDA and NMDA receptors in cochlear ischemia. Neuroreport 1992;3:299–302.
42 Pirvola U, Ylikoski J, Palgi J, Lehtonen E, Arumäe U, Saarma M: Brain-derived neurotrophic factor and neurotrophin 3 mRNAs in the peripheral target fields of developing inner ear ganglia. Proc Natl Acad Sci USA 1992;89:9915–9919.
43 Ernfors P, Lee KF, Kucera J, Jaenisch R: Lack of neurotrophin-3 leads to deficiencies in the peripheral nervous system and loss of limb proprioceptive afferents. Cell 1994;77:503–512.
44 Fariñas I, Jones KR, Backus C, Wang XY, Reichardt LF: Severe sensory and sympathetic deficits in mice lacking neurotrophin-3. Nature 1994;369:658–661.
45 Ernfors P, Duan ML, El Shamy WM, Canlon B: Protection of auditory neurons from aminoglycoside toxicity by neurotrophin-3. Nat Med 1996;2:463–467.
46 Basile AS, Huang JM, Xie C, Webster D, Berlin C, Skolnick P: N-methyl-*D*-aspartate antagonists limit aminoglycoside antibiotic-induced hearing loss. Nat Med 1996;2:1338–1343.
47 Duan M, Agerman K, Ernfors P, Canlon B: Complementary roles of neurotrophin 3 and a N-methyl-*D*-aspartate antagonist in the protection of noise and aminoglycoside-induced ototoxicity. Proc Natl Acad Sci USA 2000;97:7597–7602.
48 Staecker H, Gabaizadeh R, Federoff H, Van de Water TR: Brain-derived neurotrophic factor gene therapy prevents spiral ganglion degeneration after hair cell loss. Otolaryngol Head Neck Surg 1998;119:7–13.
49 Dulon D, Ryan AF: The bacterial Neo gene confers neomycin resistance to mammalian cochlear hair cells. Neuroreport 1999;10:1189–1193.

Allen F. Ryan, PhD, ENT, 0666, UCSD,
9500 Gilman Drive, La Jolla, CA 92093 (USA)
Tel. +1 858 534 4594, Fax +1 858 534 5319, E-Mail afryan@ucsd.edu

Felix D, Oestreicher E (eds): Rational Pharmacotherapy of the Inner Ear.
Adv Otorhinolaryngol. Basel, Karger, 2002, vol 59, pp 75–83

Pharmacological Influence on Inner Ear Endothelial Cells in Relation to the Pathogenesis of Sensorineural Hearing Loss

B. Gloddek, K. Lamm, W. Arnold

Department of Otolaryngology/Head and Neck Surgery, Technical University of Munich, Germany

Abstract

Despite an increasing incidence of acute sensorineural hearing loss, the pathogenesis of this disease remains uncertain. While viral infection of the stria vascularis, organ of Corti or spiral ganglion cells is discussed in the American literature, a vascular genesis with resulting impaired perfusion of the inner ear is favoured by European investigators. Although both hypotheses are supported by different therapeutic strategies to regain normal hearing, the influence of spontaneous remission remains unclear. This study aims at combining these seemingly opposing concepts with the assumption of an immunologically mediated vasculitis with consequent cochlear hypoperfusion.

We already know from other organs that during viral vasculitis circulating immunoglobulins are deposited perivascularly, which leads to a local decrease in perfusion and tissue hypoxia. Also in autoimmune diseases, perivasculitis is common with the endothelium playing a major role at the initial stages of the disease. These endothelial cells promote vasculitis by secreting pro-inflammatory cytokines like IL-1, IL-6 or TNF-α in addition to the expression of adhesion molecules. Due to the persistence of these immunopathological mechanisms stenosis or atresia with ischaemic necrosis results.

To examine whether this pathomechanism is also important in inner ear dysfunction, the immunological response after stimulation of the cochlear endothelium of guinea pigs was determined. In addition, the influence of corticosteroids on this immune cascade was examined.

Despite an increasing incidence of acute sensorineural hearing loss (SNHL), the pathogenesis of this disease remains uncertain. While viral infection of the stria vascularis, organ of Corti or spiral ganglion cells is discussed in the American literature [1, 2], a vascular genesis with resulting impaired perfusion of the inner ear is favoured by European investigators [3–5]. However, no correlation between altered blood fluid parameters or other cardiovascular risk factors and acute or recurrent SNHL exists [6–8]. In addition, no arteriosclerotic changed vessels could be observed in human temporal bones [9, 10].

The percentage of spontaneous remission of SNHL is estimated to be between 25 and 68% [1, 10], and therapeutic strategies with blood-flow-promoting drugs do not show any significant advantage when compared to placebo drugs [8, 11]. Reflecting these clinical data it seems important to further investigate the pathogenesis of acute SNHL in order to determine therapeutic strategies.

Concerning the viral hypothesis of SNHL, we know from animal experiments that several viral infections can affect the cochlea with resulting hearing loss [12–14]. During experimental viral labyrinthitis high endothelial venules and perivasculitis of the modiolar vessels are described [15, 16]. In studies from this laboratory, comparable changes of the cochlear vessels are observed in experimental autoimmune inner ear disease [17, 18]. In addition to the experimental data it was hypothesized that vasculitis of the labyrinth induced by viral infection might be involved in the pathogenesis of SNHL [19–21].

In the course of non-cochlear viral vasculitis, circulating immune complexes are disposed on the vessel walls [22–24] leading to local hypoperfusion and consequent tissue hypoxia. This particular pathomechanism is supposed to be responsible for panarteritis nodosa following hepatitis B viral infection or the typical vasculitis during Wegener's granulomatosis [25]. But also during autoimmune diseases, perivasculitis is a common feature with the endothelium being the main target [22, 26]. These endothelial cells (ECs) play a major role in developing vasculitis by secreting pro-inflammatory cytokines like IL-1, IL-6 and TNF-α [25, 27, 28] and expressing adhesion molecules like ICAM-1 as already demonstrated in rheumatoid arthritis, idiopathic uveoretinitis and other neuroimmunological targets [29–31]. Persistence of these immunopathological conditions will lead to stenosis or atresia of the involved vessels and necrosis of the tissue [32]. Taken together the initiation and chronification of a vasculitis is due to the imbalance between pro- and contra-inflammatory mediators secreted by ECs [33].

To examine whether this pathomechanism is also important in inner ear dysfunction, the immunological response of the cochlear endothelium obtained from guinea pigs was examined after stimulation. The spiral modiolar artery was microdissected in culture medium in order to isolate and culture ECs. After incubation with different concentrations of lipopolysaccharide (LPS) with and without fetal calf serum (FCS), the supernatant was collected and run in various assays to detect IL-1, IL-6 and TNF-α. The same experimental set-up was used to observe the effect of the synthetic glucocorticoid dexamethasone and its specific receptor antagonist RU 486.

Material and Methods

Microdissection of Cochlear Vessels

Adult pigmented guinea pigs weighing 250–350 g were decapitated, the temporal bones removed and the bulla opened. Temporal bones were then transferred into Hanks' balanced salt solution (Gibco, Bethesda, Md., USA) which was buffered with 5 m*M* sodium Hepes adjusted to pH = 7.4. All procedures were carried out at room temperature unless indicated. Under the dissection microscope, the bony wall of the cochlea was opened at the apex and removed turn by turn. The spiral modiolar artery with its primary branches was approached via the bony modiolus, separated from the modiolar tissue and transferred into a microtube containing Hanks' medium (Gibco).

Establishment of an EC Culture

The primary EC culture was established by adding specific EC medium (stimulation factor medium, SFM, Gibco) with 1% penicillin/streptomycin for 5 days until adherence to the bottom in 2% gelatine-coated T-25 flasks. The medium was changed twice every 5 days until monolayer growth. These cells were then split off the flask with 0.25% pancreatin for 3 min. The supernatant was taken off and centrifuged at 1,000 rpm for 4 min, the pellet resuspended and divided onto a 48-well plate with SFM and again cultured until monolayer growth for approximately 3 days. After formation of a complete monolayer, cells did not grow further. However, this growth arrest did not result in the induction of cell death.

Viability Test and Identification of ECs by Immunofluorescence

Cells were stained for viability testing with BCECF-AM (Molecular Probes Inc., Eugene, Oreg., USA). The percentage of viable ECs before stimulation was calculated from 5 experiments, in each counting 100 cells using a haemocytometer. In order to clearly identify ECs, an incubation with fluorescence-conjugated lectin from *Ulex europaeus* specific for ECs was performed.

Stimulation Protocol

ECs were used to study the effect of LPS after the second splitting. Cells were adapted to serum-free conditions by a stepwise reduction of FCS and SFM content. Under serum-free conditions, the growth rate was reduced and depended on the initial seeding density. LPS at concentrations between 0.1 and 1,000 ng/ml was added to the wells for 18 h in quadruplicates. 500 μl of the supernatant was taken off and immediately frozen at −80 °C. A fibroblast culture or medium without LPS served as negative control; all set-ups were run with and without 10% FCS. A mixed lymphocyte culture (MLC) was employed as positive control.

In a second experimental set-up, the effect of dexamethasone and its inhibitor RU 486 was tested. Either dexamethasone (10 ng/ml) or dexamethasone (10 ng/ml) together with RU 486 (1 mg/ml) was added simultaneously to the set-up described above and the supernatant taken off and immediately frozen at −80 °C.

Cytokine Assays (ELISA)

All samples were run in commercially available cytokine assays (R&D, Heidelberg, Germany). Since only mouse-specific assays were available, the cross-reactivity towards guinea pigs was proved by MLC. In total, three assays were employed: TNF-α with a detection minimum concentration of 5.1 pg/ml, IL-1β with minimal detection of 3.0 pg/ml and IL-6 with a detection minimum of 3.1 pg/ml. In order to exclude any contamination in the culture or reagents, all substrates were screened for TNF-α/LPS employing the established limulus lysate assay.

Results

Establishment of a Viable EC Culture

Viable ECs were obtained from the spiral modiolar artery with an average cell diameter of 2.3 μm. Single ECs exhibited fluorescence after incubation with *U. europaeus* and could by then be identified.

Control Cell Lines

An MLC as positive control known to release cytokines like IL-1, IL-6 or TNF-α proved the cross-reactivity of the employed mouse-specific assay with average values of 58 pg/ml for IL-1β, 42 pg/ml for IL-6 and 38 pg/ml for TNF-α (fig. 1). To exclude unspecific effects of co-cultivated fibroblasts in the culture, a pure fibroblast culture was run through the assays and exhibited only very low levels of cytokine after stimulation with LPS (fig. 1). Cytokine titres were below the detection level when using medium alone or unstimulated cells without LPS.

Results after Stimulation

When ECs were stimulated with LPS, the secretion of IL-1, IL-6 and TNF-α clearly depended on the addition of FCS. This secretion showed an LPS dose-dependent increase in cytokine concentrations (fig. 2a, 3a, 4a). When adding dexamethasone simultaneously to the LPS-stimulated cell culture, cytokine titres dropped dramatically to one tenth or less of the above-mentioned levels. This effect was almost completely inhibited by adding RU 486 with cytokine titres returning to high values (fig. 2b, 3b, 4b).

Discussion

In this study, cultured ECs from the spiral modiolar artery were stimulated with the unspecific inflammatory mediator LPS and were shown to secrete IL-1, IL-6 and TNF-α. This stimulation depended on the addition of FCS which

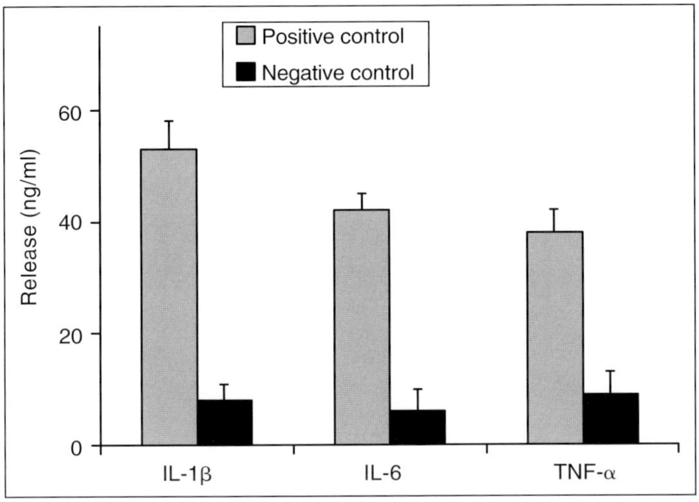

Fig. 1. Release of cytokines after LPS stimulation in MLC (positive control) and fibroblast culture (negative control), i.e. MLC or fibroblasts + 10% FCS and 100 ng/ml LPS.

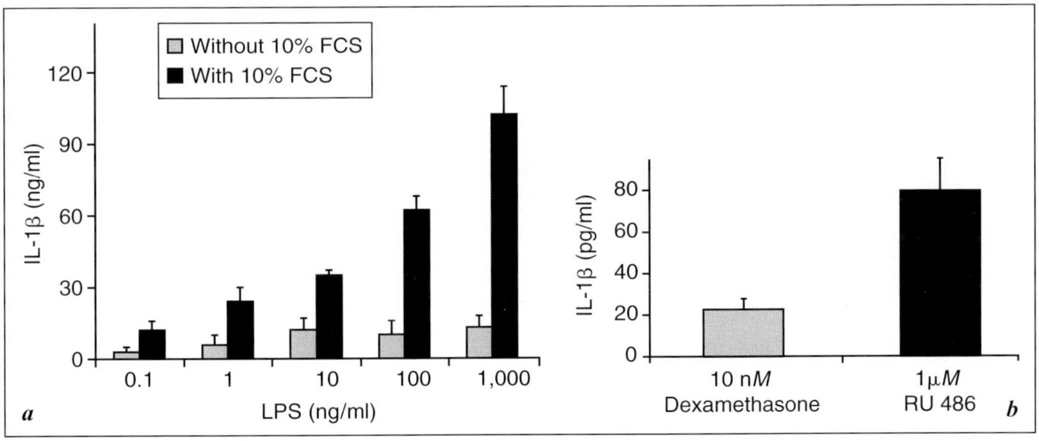

Fig. 2a, b. Release of IL-1β after LPS stimulation (***a*** 0.1–1,000, ***b*** 100 ng/ml) with/without FCS (10%) and its inhibition by dexamethasone and disinhibition by RU 486.

contains soluble CD14, mandatory to mediate the effects of LPS to the EC and functioning as a link [34].

It was demonstrated in other organ systems that the reaction of the endothelium is uniform to inflammatory agents so that it seems reasonable to use LPS to mimic the effects, for example a viral infection [35]. The cytokines detected

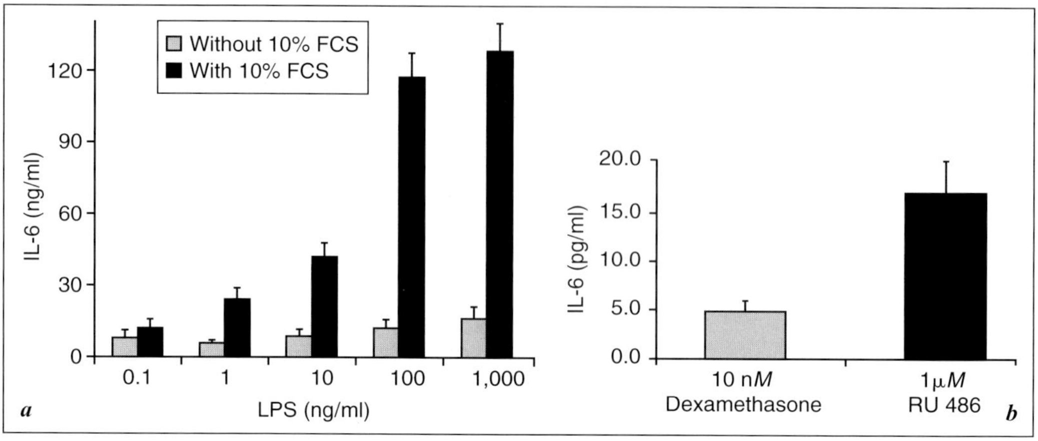

Fig. 3a, b. Release of IL-6 after LPS stimulation (*a* 0.1–1,000, *b* 100 ng/ml) with/without FCS (10%) and its inhibition by dexamethasone and disinhibition by RU 486.

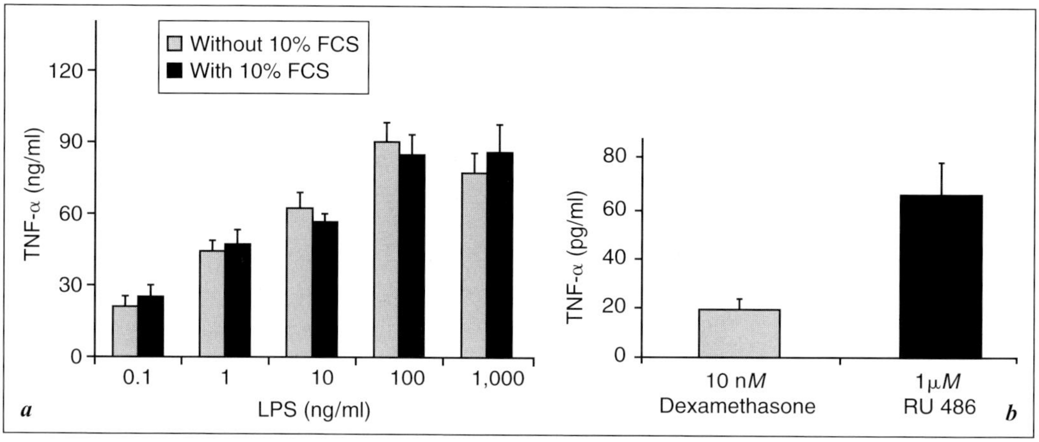

Fig. 4a, b. Release of TNF-α after LPS stimulation (*a* 0.1–1,000, *b* 100 ng/ml) of ECs with/without FCS (10%) and its inhibition by dexamethasone and disinhibition by RU 486.

above are uniformly released during viral infections and exhibit various effects during unspecific inflammation or autoimmune disease: the development of high endothelial venules and the expression of adhesion molecules, initiating vasculitis with resulting vascular leakage syndrome [36–39]. Both histopathological features are also typical in experimental labyrinthitis with the entry of immunocompetent cells into the cochlea and perivasculitis of the modiolar vessels [18]. The results are able to support our new concept of an immunologically

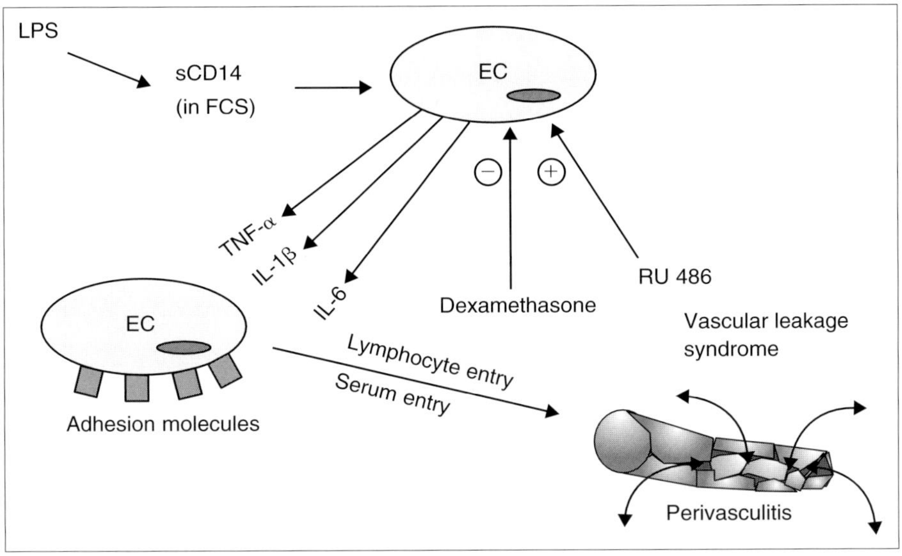

Fig. 5. Effects of ECs and their cytokine products on promoting vasculitis and vascular leakage syndrome.

mediated vasculitis as being the key factor in other diseases and gives new insights into the pathogenesis of sudden SNHL.

The ECs promote the vasculitis by secreting pro-inflammatory cytokines. Due to the persistence of these immunopathological mechanisms, stenosis or atresia with ischaemic necrosis can result.

This study combines opposing hypotheses concerning the pathogenesis of acute SNHL (virus-induced vs. vascular hypoperfusion) under the working thesis of an immunologically induced vasculitis with consequent hypoperfusion of the cochlea and structural damage. Additional studies with this model proved the therapeutic influence on the described vascular leakage syndrome (fig. 5). Dexamethasone was capable to inhibit the secretion of pro-inflammatory cytokines and its receptor antagonist RU 486 has a disinhibitory effect. Steroids seem to exhibit their effect during SNHL not only on hair cells but on cochlear vessels which are disturbed during SNHL by vasculitis. Our studies demonstrate that these cell-biological in vitro studies can lead to a better understanding of the pathological changes of the cochlear endothelium, i.e. vasculitis with resulting vascular leakage syndrome, entry of immunocompetent cells and perivasculitis. These changes should result in cochlear ischaemia, intracochlear tissue damage and hearing loss. The therapeutic influence of corticosteroids on sudden SNHL is strongly supported by our results.

Acknowledgement

This work was supported by the Deutsche Forschungsgemeinschaft DFG-La 670/1-3, Germany.

References

1 Wilson WR, Byl FM, Laird N: The efficacy of steroids in the treatment of idiopathic sudden hearing loss. Arch Otolaryngol 1980;106:772–776.
2 Schuknecht HF, Donovan ED: The pathology of idiopathic sudden sensorineural hearing loss. Arch Otolaryngol 1986;243:1–15.
3 Kellerhals B, Hippert F, Pfaltz CR: Treatment of acute acoustic trauma with low molecular weight dextran. Pract Otorhinolaryngol 1971;33:260–264.
4 Tange RA: Vascular structures of the inner ear; in Ars B (ed): Inner Ear Partition. The Hague, Kugler, 1997, pp 5–11.
5 Pirodda A, Saggese D, Ferri GG, Giusa G, Grippo MC, Gaddi A: The role of hypotension in the pathogenesis of sudden hearing loss. Audiology 1997;36:98–108.
6 Lamm K: Die medikamentöse Therapie der Innenohrschwerhörigkeiten – Kritische Anmerkungen. Wien Med Wochenschr 1992;20/21:455–459.
7 Lamm K, Arnold W: Sind durchblutungsfördernde Massnahmen (Hämorheologika) bei der Behandlung von Innenohrschwerhörigkeiten sinnvoll? Otorhinolaryngol Nova (Basel) 1993;3: 285–291.
8 Lamm K: Rationale Grundlagen einer Innenohrtherapie. Otorhinolaryngol Nova (Basel) 1995;5: 153–160.
9 Friedmann I, Arnold W: Pathology of the Ear. Edinburgh, Churchill Livingstone, 1993.
10 Schuknecht HF: Pathology of the Ear. Philadelphia, Lea & Febiger, 1993.
11 Lamm K, Arnold W: Successful treatment of noise-induced cochlear ischemia, hypoxia and hearing loss. Ann NY Acad Sci 1999;884:233–249.
12 Davis LE, Johnson RT: Experimental viral infections of the inner ear: Acute infections of the newborn hamster labyrinth. Lab Invest 1976;34:349–356.
13 Davis LE, Johnsson LG: Viral infections of the inner ear: Clinical, virologic and pathologic studies in humans and animals. Am J Otolaryngol 1983;4:347–362.
14 Woolf NK, Harris JP: Cochlear pathophysiology associated with inner ear immune responses. Acta Otolaryngol (Stockh) 1986;102:353–364.
15 Stearns GS, Keithley EM, Harris JP: Development of high endothelial venule-like characteristics in the spiral modiolar vein. Laryngoscope 1993;103:890.
16 Suzuki M, Mogi G, Harris JP: Expression of intercellular adhesion molecule-1 during inner ear inflammation. Proc 4th Conf Immunobiol Otorhinolaryngol, Oita, April 1994, pp 161–164.
17 Gloddek B, Gloddek J, Arnold W: Induction of autoimmune labyrinthitis by transfer of inner ear specific T-helper cells. Ann NY Acad Sci 1997;830:235–265.
18 Gloddek B, Lassmann S, Gloddek J, Arnold W: Role of the protein S-100 as potential autoantigen during an autoimmune disease of the inner ear. J Neuroimmunol 1999;99:451–457.
19 Gussen R: Polyarteritis nodosa and deafness: A human temporal bone study. Arch Otorhinolaryngol 1977;217:263–271.
20 Scasso CA, Bruschini L, Berrettini S, Bruschini P: Progressive sensorineural hearing loss from infectious agents. Acta Otolaryngol Ital 1998;18(suppl 59):51–54.
21 Ruckenstein MJ, McKnown KM, Jacewicz M: Unusual and instructive case of immune-mediated inner ear disease associated with central nervous system vasculitis. Otolaryngol Head Neck Surg 2000;122:109–111.
22 Abbas AK, Lichtman AH, Pober JS: Immunologie. Bern, Huber, 1996, pp 479–487.
23 Generau T, Lotholary O, Guilevin L: Physiology of viral vasculitis. Pathol Biol (Paris) 1999;47: 226–231.

24 Lortholary O, Generau T, Guillevin I: Viral vasculitis not related to HBV, HCV and HIV. Pathol Biol (Paris) 1999;47:248–251.
25 Muller Kobold AC, van Wijk RT, Franssen CF, Molema G, Kallenberg CG, Tervaert JW: In vitro up-regulation of E-selectin and induction of interleukin-6 in endothelial cells by autoantibodies in Wegner's granulomatosis and microscopic polyangiitis. Clin Exp Rheumatol 1999;17:433–440.
26 Triolo G: Enhancement of endothelial cell E-selectin expression by sera from patients with active Behçet's disease: Moderate correlation with anti-endothelial cell antibodies and serum myeloperoxidase levels. Clin Immunol 1999;91:330–337.
27 Fabry Z, Fitzsimmons KM, Herlein JA, Moninger TO, Dobbs MB, Hart MN: Production of the cytokines interleukin 1 and 6 by murine brain microvessel endothelium and smooth muscle pericytes. J Neuroimmunol 1993;47:23–34.
28 Johnson DR: Arterial and venular endothelial cell costimulation of cytokine secretion by human T cell clones. J Leukoc Biol 1998;63:612–619.
29 Dinther-Janssen CHM, Horst E, Koopman G, Newman W, Scheper WJ: The VLA/VCAM-1 pathway is involved in lymphocyte adhesion to endothelium in rheumatoid synovium. J Immunol 1991;147:4207–4210.
30 Zaman AG, Edelstein C, Stanford MR, Graham EM, Ellis BA, et al: Soluble intercellular adhesion molecule-1 (sICAM-1) as a marker of disease relapse in idiopathic uveoretinitis. Clin Exp Immunol 1994;95:60–65.
31 Dalakas MC: Basic aspects of neuroimmunology as they relate to immunotherapeutic targets: Present and future prospects. Ann Neurol 1995;37(suppl 1):2–13.
32 Wakefield TW, Strieter RM, Prince MR, Downing LJ, Greenfield LJ: Pathogenesis of venous thrombosis: A new insight. Cardiovasc Surg 1997;5:6–15.
33 Schreiber S, Stange EF: Morbus Crohn und Colitis ulcerosa: Neue Erkenntnisse zur Immunpathogenese. Dtsch Ärztebl 1997;94:1268–1274.
34 Golenbock DT, Bach RR, Lichenstein H, Juan TS-J, Tadavarthy A, Moldow CF: Soluble CD14 promotes LPS activation of CD14-deficient PNH monocytes and endothelial cells. J Lab Clin Med 1995;125:662–671.
35 Rokita H, Kupiec T, Guzik K, Koj A: Vaccinia virus-regulated acute phase cytokine production in human fibroblasts, U937 cells and endothelium. Mediators Inflamm 1998;7:73–78.
36 Maruo N, Morita I, Shirao M, Murota S: IL-6 increases endothelial permeability in vitro. Endocrinology 1992;131:710–714.
37 Neng Lai K, Leung JC, Bik Lai K, Li PK, Lai CK: Anti-DNA autoantibodies stimulate the release of interleukin-1 and interleukin-6 from endothelial cells. J Pathol 1996;178:451–457.
38 Ziesche R, Petkov V, Williams J, Zakeri SM, Mosgoller W, Knofler M, Block LH: Lipopolysaccaride and interleukin 1 augment the effects of hypoxia and inflammation in human pulmonary arterial tissue. Proc Natl Acad Sci USA 1996;29:12478–12483.
39 Alecu M, Coman G, Galatescu E: Serological level of ICAM and ELAM adhesion molecules in allergic vascularitis. Rom J Intern Med 1997;35:83–88.

Dr. Bertrand Gloddek, Department of Otolaryngology/Head and Neck Surgery,
Technical University of Munich, D–81675 Munich (Germany)
Tel. +49 89 4140 2370, Fax +49 89 4140 4893, E-Mail B.Gloddek@lrz.tum.de

Felix D, Oestreicher E (eds): Rational Pharmacotherapy of the Inner Ear.
Adv Otorhinolaryngol. Basel, Karger, 2002, vol 59, pp 84–90

Mpv37 Mouse Strain – A Model for the Relationship between the Kidney and the Inner Ear

Angela M. Meyer zum Gottesberge[a], *V. Balz*[a], *Heidi Felix*[b]

[a] Research Laboratory, Department of Otorhinolaryngology, University of Düsseldorf, Germany;
[b] Department of Otorhinolaryngology, University Hospital, Zürich, Switzerland

Abstract

Using the Mpv17-negative mouse strain, which developed inner ear and kidney dysfunction, we confirm a strong relationship between the kidney and the inner ear. Both organs have specialized epithelia involved in active ion transport, which are separated from the vessels by a basement membrane of similar composition. Our recent results indicate that the glomerular and the stria vascularis basement membrane are simultaneously affected in early stages. Concomitant deposits of IgG during the progressive development of the disease support the idea of a shared antigen. Understanding the pattern of the development of the degeneration will provide a basis towards understanding the essential role of the Mpv17 protein in the structures of both organs and may provide a basis for future therapeutic intervention.

Clinical and experimental reports established a strong relationship between the kidney and the inner ear. Both organs contain epithelia that are involved in either filtration or ion exchange (i.e. the podocytes and tubular epithelium in the kidney and the stria vascularis in the inner ear). The ototoxic and nephrotoxic effect of certain drugs was demonstrated in humans and also in experimental animals [review in ref. 1]. The fact that certain tissues of the kidney and the inner ear react immunologically in the same manner supports the concept of an autoimmuno-sensorineural etiology of hearing loss [2–5]. The kidney and the inner ear are prominent sites of type IV collagen expression, an important component of their basement membranes (BM) [6, 7]. This concept is supported by the existence of congenital anomalies related to the mutation of collagen IV [8, 9] causing lesions in both organs such as Alport syndrome [review in ref. 10].

The recessive transgenic Mpv17-negative mouse strain develops nephrotic syndrome and deafness, whereas loss of auditory function is associated with severe progressive degeneration of inner ear structures [11–13]. The Mpv17 gene product (approximately 20 kD) involved in the metabolism of reactive oxygen species (ROS) causes an imbalance of ROS homeostasis and accumulation of lipid peroxidation products [14, 15]. Northern blots of various mouse tissue mRNAs [11] and in situ analysis [16] have revealed a ubiquitous expression of Mpv17. These findings raised the question whether the inner ear phenotype in Mpv17-negative mice is a defect originating in the inner ear and in the kidney itself, or whether it is a secondary effect caused by a differently impaired metabolic function. In both organs, in the kidney and the inner ear, characteristic pathological alterations of the capillary BM are associated with the development of the disease. In this study ultrastructural observations, specific antibodies against BM compounds and detection of the immunodeposit (IgG) have been employed to investigate the BM with respect to development of the disease. Understanding the pattern of the development of degeneration provides a basis towards understanding the essential role of the Mpv17 protein in both organs. Such insights might aid in the comprehension of the nature of BM dysfunction that results in a loss of hearing and may provide a basis for future therapeutic intervention.

Material and Methods

The cochleae and the kidney from Mpv17-negative and wild-type mice from 18 days to 1 year of age were processed and evaluated either by transmission electron microscopy (TEM) or by light microscopy with immunohistochemical staining. The public authorities approved the care and use of the animals in this study.

The anesthetized animals (pentobarbiturate 60 mg/kg, i.p. together with 0.5 mg/kg atropine sulfate) were intracardially perfused with washing solution followed by fixation solution (2.5% glutaraldehyde in 0.1 *M* cacodylate buffer, pH 7.3). Alternatively, 4% paraformaldehyde in 0.1 *M* phosphate buffer, pH 7.4 or Carnoy's fixative was used. After washing in buffer the bullae were decalcified in 10% EDTA for 1–2 weeks changing the decalcifying solution daily. Cochleae for TEM were postfixed in 1% cacodylate-buffered OsO_4 for 1 h at room temperature, dehydrated in a graded series of ethanol and propylene oxide, and embedded in epoxy resin. Ultrathin sections from representative regions were stained with uranyl acetate and lead citrate and examined and photographed by a Hitachi H 7000 electron microscope.

Cochleae for immunohistochemistry were dehydrated in a graded series of ethanol and embedded in paraffin resin. 5-μm sections stained with antibodies specific for laminin B1 and B2 (Chemicon) were pretreated with 0.1 *M* glycine for 10 min and 0.1% SDS for 30 min at 60 °C. Sheep antimouse IgG conjugated with FITC (1 : 150; 1 h; room temperature) was used as a secondary antibody. Alternatively, after blocking the endogenous peroxidase with 10% H_2O_2 the sections were coated with a goat antimouse IgG (H+L) conjugated with biotin (Sigma) (1 : 100; 3 h; room temperature), washed with PBS, and flooded with streptavidin-avidin-horseradish peroxidase complex and a substrate 3,3-DAB as recommended when using the Vectastain kit (Vector Laboratories).

Results and Discussion

Early stages of the disease showed glomerular lesions related to a small number of glomeruli. BM showed thinning, thickening or splitting and formation of a 'basket weaving' at the age of 18 days (fig. 1a). TEM studies revealed degenerative changes in the visceral epithelial cells associated with fusion of the foot processes. The subendothelial space of capillaries contained homogeneous electron-dense material with foamy structures. In the stria vascularis at the age of 18 days the BM displayed clear multiple laminations (fig. 2a–c). The lesion progressed first with moderate thickening of the BM followed by a significant

Fig. 1. Transmission electron micrographs of the glomerular BM of Mpv17-negative mice. ***a*** 18 day-old-mouse showing lamination (open arrows) and basket-weaving structure characteristic of Alport's syndrome. ***b*** Enlarged and sclerotic BM of 7-month-old mouse. Bar: 1 μm.

Fig. 2. Transmission electron micrographs of the capillary BM in the stria vascularis of Mpv17-negative mice. ***a*** 18-day-old mouse showing lamination of the BM (double arrows). ***b*** 2-Month-old mouse with moderate enlargement of the BM and foamy structures (arrow) occasionally present in enlarged BM. ***c*** Old Mpv17-negative mouse with foamy structures frequently inserted in BM. Bar: 1 μm.

Fig. 3. Immunohistochemical staining for laminin B2 in the stria vascularis of a wild-type (***a***) and a 6-month-old Mpv17-negative mouse (***b***). × 240.

Fig. 4. Immunohistochemical staining for laminin B1 in the stria vascularis of a wild-type (***a***) and 6-month-old Mpv17-negative mouse (***b***). Arrowheads indicate the location of the stria vascularis. × 240.

Fig. 5. Immunohistochemical staining for laminin B2 in the kidney of a wild-type (***a***) and a 6-month-old Mpv17-negative mouse (***b***). × 240.

Fig. 6. Immunohistochemical staining for laminin B1 in the kidney of a wild-type (***a***) and a 6-month-old Mpv17-negative mouse (***b***). × 240.

Fig. 7. Immunohistochemical staining for IgG deposit in the stria vascularis of a wild-type (***a***), a 2-month-old (***b***) and a 7-month-old Mpv17-negative mouse (***c***). Arrowheads indicate the location of the stria vascularis. × 400.

Fig. 8. Immunohistochemical staining for IgG deposit in the kidney of a wild-type (***a***), a 2-month-old (***b***) and a 7-month-old Mpv17-negative mouse (***c***). (*) indicate location of glomeruli. × 250.

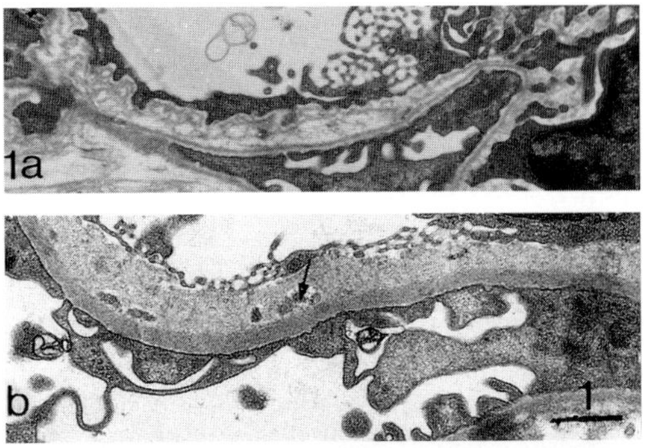
1a
b
1

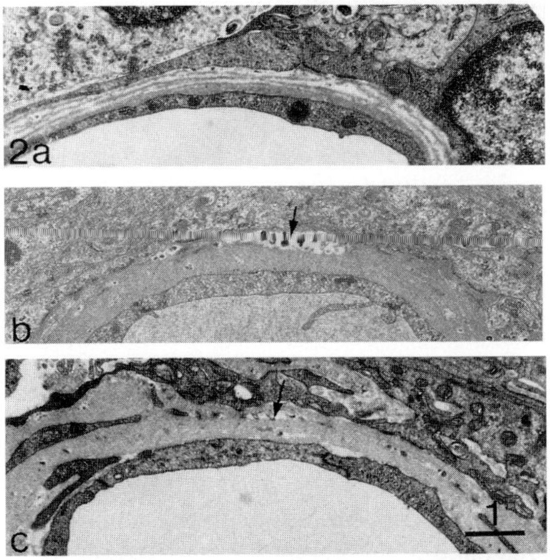
2a
b
c
1

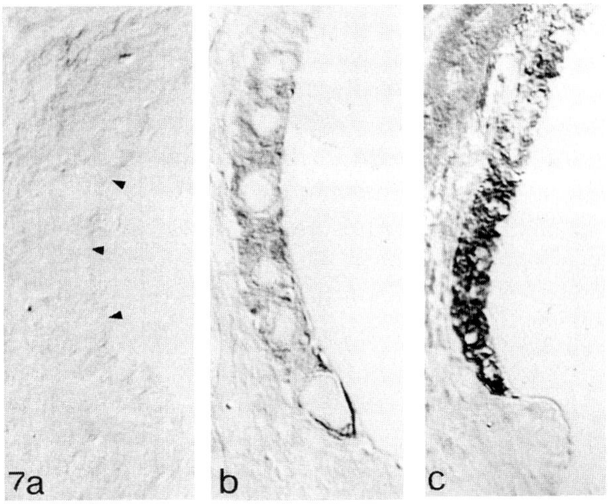
7a
b
c

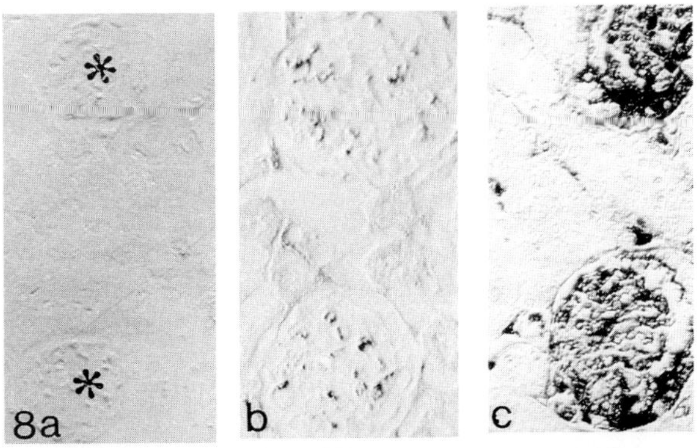
*
*
8a
b
c

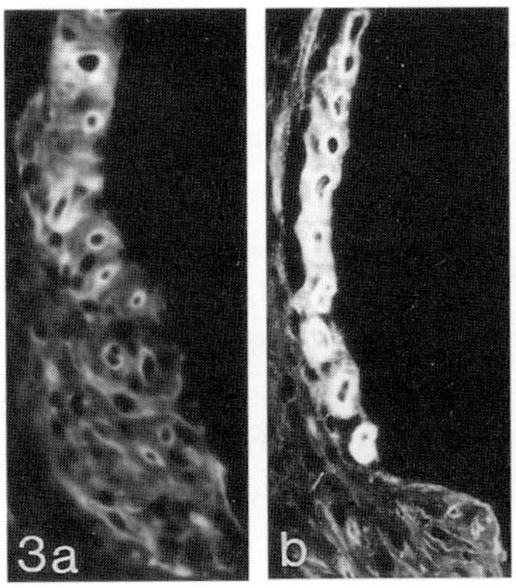
3a
b

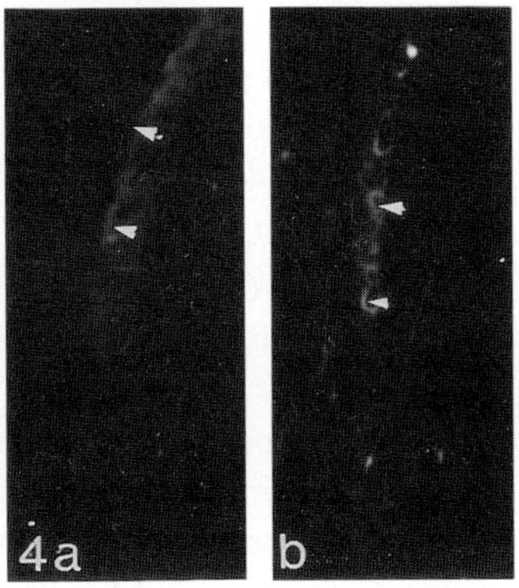
4a
b

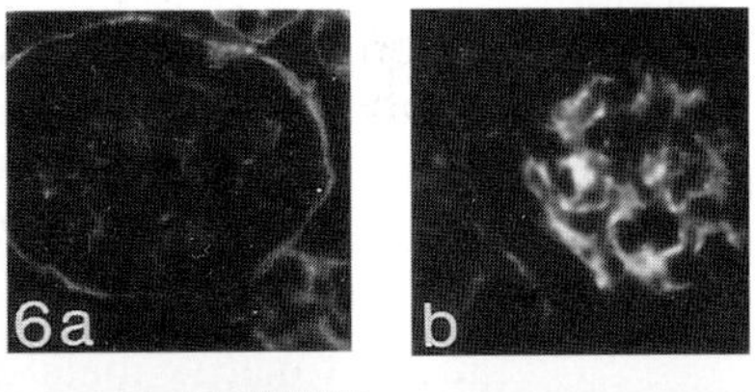
6a
b

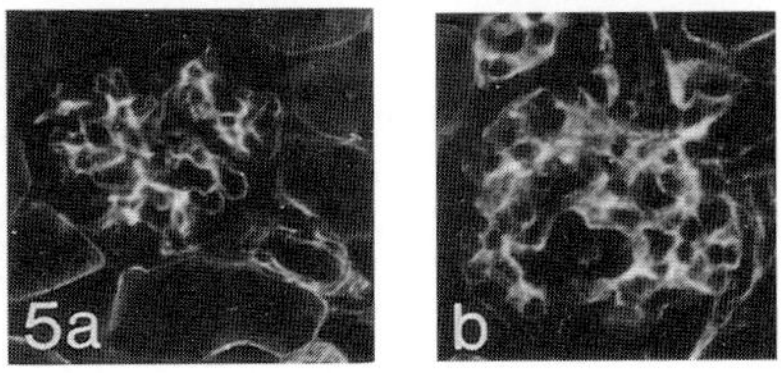
5a
b

enlargement. In the expanded BM deposit of amorphous particles of varying electron density and electronlucent areas enclosing dense particles (foamy-like structures) were present. Comparing the development of their structural alterations in the BM, it is obvious that these alterations are concomitant in both organs and that similar intermediate signals are most probably involved in the development of the disease.

In Mpv17-negative mice structural changes of the BM of the capillary in the stria vascularis and the glomeruli indicated numerous changes in the molecular composition. Defective distribution of the collagen IV subunits (results not shown) and accumulation of the B2 laminin in the inner ear tissue were found. By contrast, in the glomeruli accumulation of laminin B1 was detected (fig. 3–6).

Moderate strial atrophy and thickening of the capillary basement membranes were reported under a variety of pathological conditions including immune processes with or without kidney involvement and aging [7, 17, 18]. The presence of foamy-like structures may be related to the immunodeposit; therefore, the presence of immunoglobulins in the thickened BM was investigated immunohistochemically. We found accumulation of IgG in the stria vascularis (fig. 7b, c) and in the glomeruli (fig. 8b, c) which started at about 2 months of age. Immunodeposit was coexistent in both organs (fig. 7, 8), but occurred later than ultrastructural alterations. We conclude that if an 'autoimmune process' plays a role here, it is a secondary effect that is not present at disease onset.

The otopathology seen in Mpv17 mutants, especially the alteration of the capillary BM closely resembles electron microscopic observations of the inner ear tissue in patients with Alport syndrome [19] and in collagen IV α3-knockout mice [20]. Alport syndrome is predominantly caused by a mutation of genes coding for the α3, α4 and α5 chains of IV collagen [8, 9]. By contrast, the primary cause of the disease in animals investigated here is the loss of function of the peroxisomal Mpv17 protein that contributes to the intracellular redox status [14]. Recently we found that the Mpv17 protein plays a crucial role in the regulation of matrix metalloproteinase-2 [16], a member of the large subfamily of proteinases, which play a major role in the physiological control of extracellular matrix components.

The Mpv17 gene has originally been described as a glomerulosclerosis gene. However, our results illustrated that the Mpv17 gene product is essential for the inner ear and that similar intermediate signals may play a crucial role in the development of the disease. Furthermore, Mpv17-negative mice may serve as a valuable model to study the molecular mechanism of this renal/inner ear disorder.

Acknowledgements

The authors wish to thank Mrs. V. Hoffmann (Zürich) and U. Becker-Lendzian (Düsseldorf) for their excellent technical help, Mrs. A. Trenker and C. Schettl (Düsseldorf) for their care of the animals. Special thanks have to be addressed to PD Dr. H. Weiher for the Mpv17-negative mice strain and to Prof. P. Wangemann and Prof. J. Wood for critical reading of the manuscript. This study was supported by Deutsche Forschungsgemeinschaft (ME 890/4-1).

References

1 Garetz SL, Schacht J: Ototoxicity: Of mice and man; in Van De Water TR, Popper AN, Fay RF (eds): Clinical Aspects of Hearing. New York, Springer, 1996, pp 116–154.
2 Quick CA, Fish A, Brown C: The relationship between cochlea and kidney. Laryngoscope 1973; 83:1469–1482.
3 Arnold W, Weidauer H, Seelig HP: Experimenteller Beweis einer gemeinsamen Antigenizität zwischen Innenohr und Niere. Arch Otorhinolaryngol 1976; 212:99–117.
4 McCabe BF: Autoimmune sensorineural hearing loss. Ann Otol Rhinol Laryngol 1979;88: 585–589.
5 Tago C, Yanagita N: Cochlear and renal pathology in the autoimmune strain mouse. Ann Otol Rhinol Laryngol 1992;101:87–91.
6 Tokahashi M, Hokunan K: Localization of type IV collagen and laminin in the guinea pig inner ear. Ann Otol Rhinol Laryngol 1992;191:58–62.
7 Cosgrove D, Samuelson G, Pitt J: Immunohistochemical localization of basement membrane collagens and associated proteins in the murine cochlea. Hear Res 1996;97:54–65.
8 Barker DF, Hostikka SL, Zhou J, Chow LT, Oliphant AR, Gerken SC, Gregory MC, Skolnick NH, Atkin CL, Tryggvason K: Identification of mutation in the COL4A5 collagen gene in Alport syndrome. Science 1990;248:1224–1227.
9 Mochizuki T, Lemmink HH, Mariyama M, Antignac C, Gubler MC, Pirson Y, Verellen-Dumoulin Ch, Chan B, Schröder CH, Smeets HJ, Reeders ST: Identification of mutations in the $\alpha3$(IV) and $\alpha4$(IV) collagen genes in autosomal recessive Alport syndrome. Nature Genet 1994; 8:77–82.
10 Friedmann I, Arnold W: Pathology of the inner ear. London, Churchill Livingstone, 1992.
11 Weiher H, Noda T, Gray DA, Sharpe AH, Jaenisch R: Transgenic mouse model of kidney disease: Insertional inactivation of ubiquitously expressed gene leads to nephrotic syndrome. Cell 1990;62: 425–434.
12 Müller M, Smolders JWT, Meyer zum Gottesberge AM, Reuter A, Zwacka RM, Weiher H, Klinke R: Loss of auditory function in transgenic Mpv17-deficient mice. Hear Res 1997;114: 259–263.
13 Meyer zum Gottesberge AM, Reuter A, Weiher H: Inner ear defect similar to Alport's syndrome in the glomerulosclerosis mouse model Mpv17. Eur Arch Otorhinolaryngol 1996;253:470–474.
14 Zwacka R, Reuter A, Pfaff E, Moll J, Gorgas K, Karasawa M, Weiher H: The glomerulosclerosis gene Mpv17 encodes a peroxisomal protein producing reactive oxygen species. EMBO J 1994; 13:5129–5134.
15 Binder CJ, Weiher H, Exner M, Kerjaschki D: Glomerular overproduction of oxygen radicals in Mpv17 gene-inactivated mice causes podocyte foot processes flattening and proteinuria. J Pathol 1999;154:1067–1075.
16 Reuter A, Nestl A, Zwacka RM, Tuckermann J, Waldherr R, Wagner EM, Höyhtyä, M, Meyer zum Gottesberge AM, Angel P, Weiher H: Expression of the recessive glomerulosclerosis gene Mpv17 regulates MMP-2 expression in fibroblasts, the kidney and the inner ear. Mol Biol Cell 1998;9: 1675–1682.

17 Thomopoulos GN, Spicer SS, Gratton MA, Schulte BA: Age-related thickening of basement membrane in stria vascularis capillaries. Hear Res 1997;111:31–41.
18 Sakagutchi N, Spicer SS, Thomopoulos GN, Schulte BA: Immunoglobulin deposition in thickened basement membranes of aging strial capillaries. Hear Res 1997;109:83–91.
19 Weidauer H, Arnold A: Strukturelle Veränderungen am Hörorgan beim Alport-Syndrom. Z Laryngol Rhinol Otol 1976;55:6–16.
20 Cosgrove D, Samuelson G, Meehan DT, Miller C, McGee J, Walsh EJ, Siegel M: Ultrastructural, physiological and molecular defects in the inner ear of a gene-knockout mouse model for autosomal Alport syndrome. Hear Res 1998;121:84–98.

Prof. Dr. A.-M. Meyer zum Gottesberge,
Research Laboratory, Department of Otorhinolaryngology,
Heinrich-Heine-University of Düsseldorf, Moorenstrasse 5 D–40225 Düsseldorf (Germany)
Tel. +49 211 8118247, Fax +49 211 811 9794, E-Mail gottesbe@uni-duesseldorf.de

Felix D, Oestreicher E (eds): Rational Pharmacotherapy of the Inner Ear.
Adv Otorhinolaryngol. Basel, Karger, 2002, vol 59, pp 91–95

Sensitivity to Glutamate Neurotoxicity in Different Developmental Periods of the Rat Cochlea

F. Carricondo, M.V. Bartolomé, M.A. Vicente-Torres, P. Fernández-Pacheco, T. Rodríguez, P. Gil-Loyzaga

Agustín Bullón Cell Culture Center and Department of Surgery II (ORL), Faculty of Medicine, University Complutense, Madrid, Spain

Abstract

Cochlear neurotoxicity induced by the intraperitoneal administration of monosodium glutamate (MSG) has been analyzed during the postnatal development of the auditory receptor of the rat. The animals were treated with MSG during two postnatal periods. The electrophysiological recordings showed that MSG treatment produced a decrease in the 8th nerve compound action potential. The effect was more marked in the animals treated between the 9th and 12th postnatal day than in the others, with a qualitative decrease in neuronal density in the spiral ganglion. These results suggest that there is a period of maximum sensitivity to the cochlear neurotoxicity induced by MSG in the postnatal development of the rat.

Glutamate has been identified as the main excitatory neurotransmitter of the nervous system [1]. In hypoxic conditions, large amounts of glutamate are released to the synapsis where they induce a neurotoxic process [2] which leads to the death of glutamatergic neurons, both in the central and peripheral nervous system.

In the cochlea, hypoxia or acoustia trauma produces typical neurotoxicity damage in the afferent type I neurons of the spiral ganglion [3]. Damage, such as swelling of the dendritic terminals and death of these neurons and electrophysiological changes in the cochlear potentials, was studied using treatment with glutamatergic agonists [4–11].

There is evidence that demonstrates early expression of the glutamatergic receptors at the dendritic terminals of the afferent type I fibers. It was described that the auditory receptor is sensitive to the neurotoxic action of kainic acid on the 17th gestational day [6] and on the 6th postnatal day [4] in the rat. In addition,

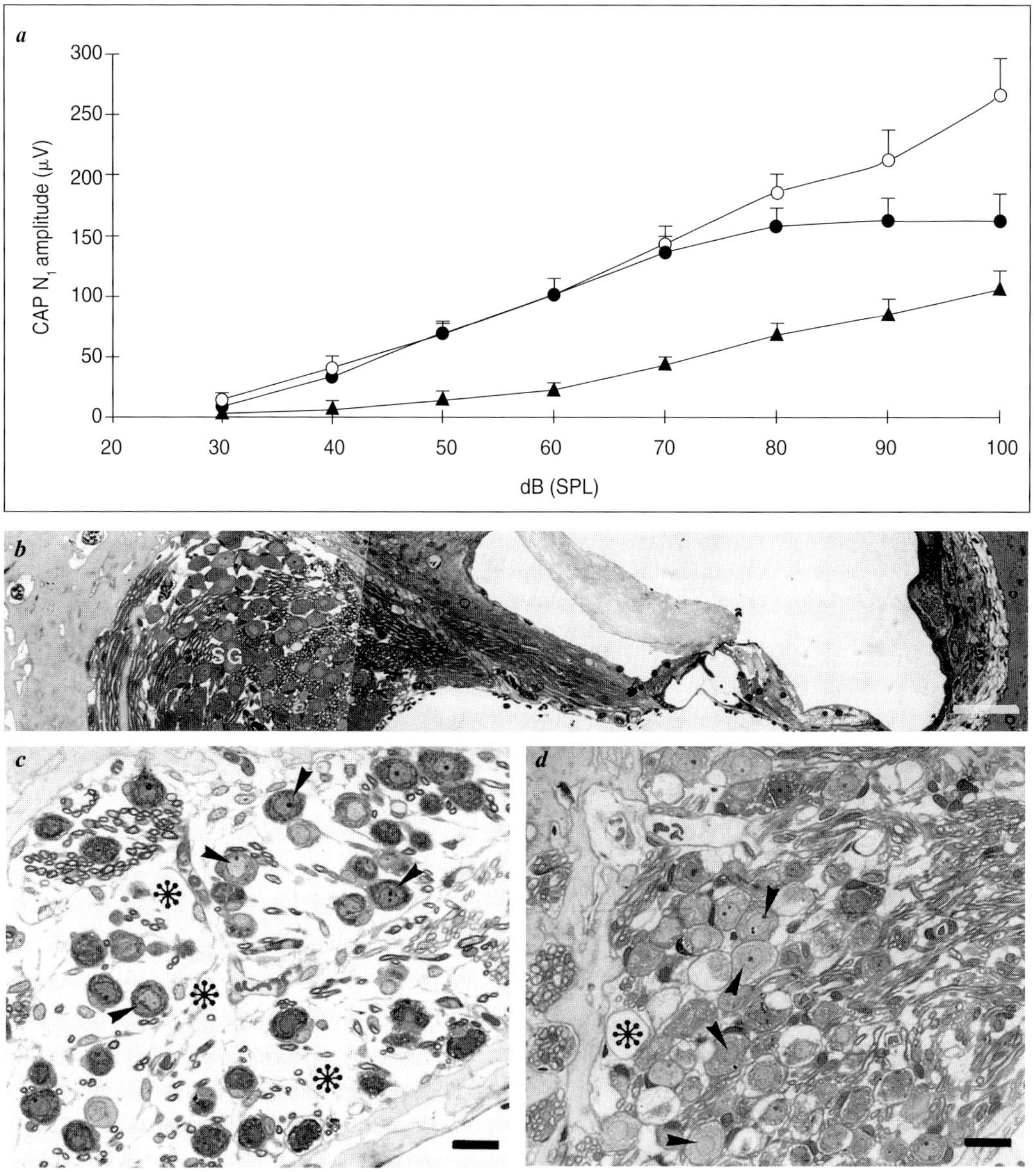

rats treated with MSG between the 2nd and 9th postnatal day showed a significant loss of the spiral ganglion cells [9].

The aim of this study was to explore the neurotoxic effects of MSG treatment during the cochlear postnatal development of the rat in order to detect periods sensitive to neurotoxicity.

Material and Methods

Long-Evans male rats were treated with MSG during postnatal development. They were distributed into two treatment groups: postnatal days 9–12 and 30–33. The control group was treated with saline. The experimental groups received a daily intraperitoneal injection of MSG (4 g/kg body weight) during the 4 days of each period.

When the animals reached postnatal day 45, they were anesthesized to record the compound action potential (CAP) of the auditory nerve by electrocochleography using clicks (10 ms) with intensities between 100 and 30 dB (SPL) as stimuli (Mistral, Medelec Int.). Then the cochlear morphological study was carried out as previously described [11].

Results

Electrophysiological Results

The amplitudes of the CAP N_1 wave differed significantly between treatment groups. The lowest amplitude was detected when MSG treatment was carried out during postnatal days 9–12. The amplitudes measured in the animal treated during postnatal days 30–33 were the highest among the treated animals, but were lower than in controls (fig. 1a).

Morphological Results

Treatment with MSG did not significantly affect the normal morphology of the organ of Corti. During postnatal days 9–12 (fig. 1c), MSG induced a

Fig. 1. Comparative analysis of the electrophysiological and morphological effects of MSG treatment in the rat postnatal development. ***a*** Input-output functions of the CAP N_1 wave amplitude. Means + SEM are shown at each level of stimulation for each of the three treatment groups. ○ = Controls; ▲ = MSG, 9–12 postnatal days; ● = MSG, 30–33 postnatal days. ***b*** Low magnification of the organ of Corti and the spiral ganglion (SG) of a control animal. ***c*** Medium magnification of the spiral ganglion of an animal treated with MSG during postnatal days 9–12. ***d*** Spiral ganglion of an animal treated with MSG during postnatal days 30–33. A significant decrease in neurons (arrowheads) was typically observed comparing an animal treated with MSG during postnatal days 9–12 (***c***) with a control animal (***b***). A few hollows (asterisks) can be observed in the cochlea of an animal treated with MSG on postnatal days 30–33 (***d***). Scale bar = 25 μm for ***b*** and 10 μm for ***c*** and ***d***.

qualitative decrease in the neuronal density of the cochlear spiral ganglion compared to controls (fig. 1b). Images obtained after MSG treatment during postnatal days 30–33 (fig. 1d) did not show any qualitative neural density reduction in the spiral ganglion.

Discussion

The present results showed a significant loss of spiral ganglion neurons in the cochleae of animals treated with MSG during postnatal days 9–12, together with a decrease in the amplitude of the CAP N_1 wave compared to controls and the recordings of the animals treated during postnatal days 30–33. These electrophysiological and morphological findings fit well since CAP reflects the synchronous discharge of the spiral ganglion neurons and the number of neurons that form the potential [7].

The present results are also in agreement with previous studies analyzing the neurotoxic process induced by MSG [8, 9] or any other glutamatergic agonist studied [4–7, 9–11]. They suggest that there is a developmental period in which the organ of Corti is more sensitive to MSG-induced neurotoxic damage. Developmentally sensitive periods were described for other ototoxic drugs (e.g. aminoglycosides) [12]. This period, i.e. the period described in this experiment, could be related with the latest developmental stages, in which hair cell differentation and synaptogenesis occur.

Acknowledgement

This work was supported by grants PM 990054 and FIS 98/0732.

References

1 Nicholls DG: The glutamatergic nerve terminal. Eur J Biochem 1993;212:613–631.
2 Olney JW: Brain lesions, obesity and other disturbances in mice treated with monosodium glutamate. Science 1969;164:719–721.
3 Puel JL, Pujol R: Selective glutamate antagonist block the excitotoxicity caused either by glutamate agonists or by ischemia; in Cazals Y, Demany L, Horner KC (eds): Advances in Biosciences: Auditory Physiology and Perception. Oxford, Pergamon Press, 1992, pp 589–598.
4 Pujol R, Lenoir M, Robertson D, Eybalin M, Johnstone BM: Kainic acid selectively alters auditory dendrites connected with cochlear inner hair cells. Hear Res 1985;18:145–151.
5 Juiz JM, Rueda J, Merchán JA, Sala MA: The effects of kainic acid on the cochlear ganglion of the rat. Hear Res 1989;40:65–74.
6 Gil-Loyzaga P, Pujol R: Neurotoxicity of kainic acid in the rat cochlea during early developmental stages. Eur Arch Oto-Rhino-Laryngol 1990;248:40–48.

7 Puel JL, Pujol R, Ladrech S, Eybalin M: α-Amino-3-methyl-4-isoxazole propionic acid electrophysiological and neurotoxic effects in the guinea pig cochlea. Neuroscience 1991;45:63–72.
8 Janssen R: Glutamate neurotoxicity in the developing rat cochlea is antagonized by kynurenic acid and MK-801. Brain Res 1992;590:201–206.
9 Janssen R, Schweitzer L, Jensen KF: Glutamate neurotoxicity in the developing rat cochlea: Physiological and morphological approaches. Brain Res 1991;552:255–264.
10 Gil-Loyzaga P, Hernández-Ortiz MJ, Rodríguez-Benito T, Lasso de la Vega M: Diltiazem protects against neurotoxicity induced by neurotoxic amino acids on cochlear afferent fibers. A morphological study. ORL 1993;55:211–215.
11 Gil-Loyzaga P, Hernández E, Carricondo F, Simón F, Poch-Broto J: Trimetazidine prevents cochlear lesions induced by intraperitoneal and perilymphatic administration of kainic acid. Brain Res 1999;826:95–103.
12 Pujol R: Periods of sensitivity to antibiotic treatment. Acta Otolaryngol Suppl 1986;429:29–33.

Prof. P. Gil-Loyzaga, Apartado de Correos 60075, E-28080 Madrid (Spain)
Tel. +34 91 3941375, Fax +34 913941383, E-Mail loyzaga@eucmax.sim.ucm.es

Felix D, Oestreicher E (eds): Rational Pharmacotherapy of the Inner Ear.
Adv Otorhinolaryngol. Basel, Karger, 2002, vol 59, pp 96–105

Protective Mechanisms of Sound Conditioning

Xianzhi Niu, Barbara Canlon

Department of Physiology and Pharmacology, Karolinska Institutet, Stockholm, Sweden

Abstract

Evidence continues to accumulate demonstrating the importance of reducing the deleterious effects of noise trauma by sound conditioning. Sound conditioning is an active process induced by low-level, nondamaging noise exposure that creates long-term protective effects to subsequent detrimental forms of noise trauma. This phenomenon is now shown to occur in a variety of mammals, including gerbils, chinchillas, guinea pigs, rabbits, rats, mice and human subjects. Different sound-conditioning paradigms have been proven successful in preventing pathological changes to the auditory system. These studies are reviewed in the present chapter and the possible biological mechanisms underlying this phenomenon are discussed.

A number of recent studies have shown that the susceptibility of the inner ear to noise trauma can be reduced by prior exposure to an acoustic stimulus. The existence of the resistance to noise trauma was first suggested by Miller et al. [1] on the basis of experiments in cats. These authors demonstrated that when cats were exposed to interrupted noise for 16 continuous days the threshold shifts declined during the latter part of the exposure compared to the thresholds obtained on the first day. Today, two distinct paradigms are employed to reduce the susceptibility of the inner ear to noise trauma. The first uses a low-level, nondamaging continuous acoustic stimulus before the traumatic exposure. This phenomenon has been termed 'sound conditioning' and has been demonstrated on a number of species including guinea pigs, gerbils, rabbits and rats [2–9]. The second paradigm uses an interrupted schedule at sound levels that produce a temporary threshold shift during the first few days of exposure. However, as the daily exposure continues, the degree of threshold shift is reduced, in some cases to no threshold shift despite an ongoing exposure. This reduction has been termed 'toughening' or resistance to noise-induced hearing loss (NIHL). Toughening has been demonstrated in chinchillas, guinea pigs and gerbils [9–21]. In low-level, continuous-conditioning studies, several investigators have reported that the conditioning stimulus should not cause significant temporary or permanent threshold shifts or hair cell damage, in order to be maximally effective in preventing subsequent hearing loss and hair cell damage. In some cases, this has been accomplished by using a relatively low level of intensity for the conditioning stimulus [2]. In other cases, a period of 'rest' has been interposed between the conditioning and damaging stimuli, in order for thresholds to recover to preconditioning levels [3].

It is important that the sound-conditioning parameters are correctly chosen. A good example of choosing the wrong stimulus parameters for sound conditioning is exemplified in a study by Fowler et al. [22]. In this study, the mouse was chosen as the experimental animal to determine the effect of sound conditioning on a subsequent high-intensity noise exposure. The mice were either conditioned to a continuous sound conditioner or to an interrupted paradigm. No protective effect was demonstrated with either paradigm. Mice appeared particularly resistant to the highly traumatic noise exposure (12 or 24 h) whereas they were particularly sensitive to the continuous sound-conditioning stimulus, which induced a threshold shift. This is intriguing when one considers the relatively low intensities used in the continuous training (threshold shifts induced) compared to either the interval training (threshold shifts not induced) or the traumatic exposure. These results cannot be explained on the basis of energy. For example, the interval training (96 dB SPL, 6 h/day for 10 days) resulted in nearly 4 times the total energy of the animals trained continuously at 80 dB SPL for 24 days. Obviously, hearing loss was not related to total acoustic energy in any simple manner. Employing equal energy may also not be an appropriate method for comparing the effects of interrupted and continuous conditioning paradigms. The lack of conditioning in the mouse might be a peculiarity of this species. However, it seems more parsimonious to conclude that the stimuli used in this particular study were inappropriate.

Recently, Yoshida and Liberman [23] demonstrated the protective effect of sound conditioning against subsequent noise trauma in mice. These investigators used two different conditioning paradigms, i.e. one of 1 week duration and the other of 15 min duration. After both sound-conditioning protocols, increased amplitudes of distortion product otoacoustic emissions were found. Both sound-conditioning paradigms resulted in reduction of noise-induced

Table 1. Different paradigms used to protect against trauma by preconditioning

Species	Condition	Pause	Trauma	Reference
Chinchilla	OBN 0.5 kHz, 95 dB, 6 h/day 10 days	5 days	Impulse noise, 150 dB	19
Chinchilla	OBN 0.5 kHz, 95 dB, 6 h/day 10 days	max 60 days	OBN 0.5 kHz, 106 dB, 48 h	21
Gerbil	OBN (1,414–5,656 Hz)	max 3 weeks	OBN (1.4–5.6 Hz) 110 dB, 1 h	3
Gerbil	OBN at 2 kHz, 74 dB, 10 days	2 days	OBN at 2 kHz, 107 dB, 48 h	9
Gerbil	OBN at 2 kHz, 80 dB, 6 h/day 10 days	2 h	OBN at 2 kHz, 107 dB, 48 h	9
Guinea pig	1 kHz, 81 dB, 24 days	none	1 kHz, 105 dB, 72 h	2
Guinea pig	6.3 kHz, 78 dB, 13 days	none	6.3 kHz, 100 dB, 24 h	7
Guinea pig	BBN, 85 dB, 5 h/day 10 days	5 days	2–20 kHz, 110 dB, 5 h	6
Rat	OBN at 4 kHz, 55–95 dB, 10 h	10 h	OBN at 4 kHz, 105 dB, 13 h	8
Rabbit	OBN at 1 kHz, 95 dB, 3 weeks		4.215 kHz, 95 dB, 5 min	13
Mouse	OBN (8–16 kHz), 89 dB, 15 min	24 h	OBN (8–16 kHz) 100 dB, 2 h	23
Mouse	heat stress (41.5 °C)	6 h	OBN (8–16 kHz) 100 dB, 2 h	25
Mouse	BBN (4–25 kHz), 70 dB, 12 h		hereditary hearing loss	24
Human	music, 70 dBA, 6 h/day 5 days	none	OBN 2 kHz, 105 dB, 10 min	18

OBN = Octave band noise.

permanent threshold shift from a subsequent high-level exposure. These findings are not in contradiction with the study by Fowler et al. [22] different sound-conditioning parameters were used. These two studies emphasize the importance of selecting suitable parameters. If protection is not found in a given condition, it is important that the results be interpreted with caution. One obvious explanation would be that optimal sound-conditioning parameters were not tested.

It has been reported that low-level acoustic stimulation could slow, but not prevent, genetically determined hearing loss in mice [24]. In order to delay the hearing loss it was important to initiate the low-level stimulation before the onset of hearing loss. These findings expand the possibilities of protecting against hearing loss by sound conditioning such that pretreatment can also protect against hereditary hearing loss. Another interesting finding regarding sound conditioning in the mouse was recently demonstrated [23]. The conditioner used in this case was whole-body heat stress. When the mice were primed with heat stress and then subjected to noise trauma a protection against hearing loss was evident compared to the group not heat stressed [25]. Both these findings in mice are suggesting that sound conditioning may have a wider range of applications in preserving hearing than previously thought. Table 1 illustrates some of the different paradigms that have been used to protect against trauma by preconditioning.

Possible Mechanisms Underlying the Phenomenon of Sound Conditioning

Many hypotheses have been advanced to explain the protective effects of sound conditioning, but the vast majority of studies are inconclusive. There is increasing evidence for endogenous protective systems in the cochlea, which if enhanced, can provide protection against subsequent trauma. Endogenous cochlear protective systems characterized to date include endogenous antioxidants or free radical scavengers, calcium-buffering systems, heat shock proteins (HSPs), glutamate receptors, and neurotrophic factors.

Endogenous Antioxidants

Following noise exposure, free radical levels have been shown to increase within the inner ear [26, 27], suggesting that NIHL reflects, at least in part, oxidative damage. In the absence of intervention, the ear apparently upregulates endogenous glutathione expression to scavenge free radicals following noise [28] and alterations in glutathione levels alter susceptibility to NIHL [29]. Enhancing free radical scavenging systems through the treatment of mannitol and deferoxamine mesylate [30], *R*-phenylisopropyladenosine [31], allopurinol or superoxide dismutase [32] protection against hearing loss and hair cell loss in animals exposed to noise has been provided. Recent studies in the chinchilla have demonstrated changes in the expression of endogenous free radical scavengers with sound conditioning [33], suggesting that the modulation of these systems may be one of the contributions of sound conditioning.

A local or generalized stress response induced by sound conditioning could result in increased levels of antioxidant enzymes in the cochlea. It was found that the combined treatment of sound conditioning and noise caused significant increases in the concentration of antioxidant enzymes compared to the group exposed to noise group [33]. Conditioning exposures may protect hair cells by increasing activities of some antioxidant enzymes, such as glutamylcysteinyl synthase, and catalase in the cochlea. Hair cells in the organ of Corti are protected from noise-induced damage by increasing stria vascularis levels of catalase, a hydrogen-peroxide-scavenging enzyme, and of enzymes involved in maintaining glutathione in the reduced state. The model formulated by these hypotheses suggests that agents that protect or augment the glutathione system in the cochlea may be protective. Reactive oxygen species (ROS) generation and an increase in Ca^{2+} are considered to be two main streams of damage leading to hair cell death [34]. Endogenous antioxidants such as glutathione are presumed to protect hair cells by scavenging ROS [29, 34].

Excess release of excitatory amino acids (excitotoxicity) may eventually lead to ROS generation and increased Ca^{2+} [35]. Nitric oxide (NO) is a typical retrograde signal in the brain, and the concentration at which it is present determines whether it has protective or toxic effects. Furthermore, NO has been shown to be one of the underlying molecules involved in the cell death following N-methyl-*D*-aspartate (NMDA)-mediated excitotic damage in the central nervous system. It is possible that NO is a mediator of aminoglycoside-induced cell damage also in the cochlea because: (1) at high concentrations NO causes the death of any type of cell (thus both hair cells and supporting cells would succumb to it), (2) it is made in spiral ganglion neurons (3) it is known that excess stimulation of NMDA receptors leads to excess NO release (4) blocking NO synthesis in the cochlea with the inhibitor N^G-methyl-*L*-arginine prevents chemically induced cytotoxicity. As previously mentioned, the overproduction of NO would be expected to damage all cell types in the organ of Corti. The NO hypothesis would also explain the loss of both hair cells and supporting cells following noise damage. Any toxic effect from the neurons to the organ of Corti cells are most likely in balance with protective mechanisms also stemming from the spiral ganglion neurons.

Calcium-Buffering Systems

Disruption of actin filaments in the hair cells by sound conditioning may act to attenuate hair cell damage induced by noise trauma. F-actin may increase the mechanical rigidity of fracture or ripping of tight cell junctions by noise trauma [36]. Sound conditioning protects hair cells by cytochalasins, compounds that disrupt actin filament by suppressing glutamate-induced Ca^{2+} influx. Canlon et al. [7] have shown that calbindin D (28 kD) immunoreactivity in the outer hair cells is decreased in lower third turn through the upper basal turn following sound conditioning in guinea pigs. Upregulating the calcium-buffering system, such as calbindin, calsequestrin, and parvalbumin, responsible for maintaining low levels of Ca^{2+} by sound conditioning is an important pathway to protect hair cells.

Heat Shock Proteins

HSPs are known to have protective effects on cells in a variety of biological systems [37, 38]. HSPs have been related to the acquired tolerance and have therefore been suggested to be involved in protection against various types of insults. In normal unstressed guinea pig cochlea there is a constitutive level of HSP 72 expression in Deiters cells and interdental cells of the spiral limbus [39], and the expression has been shown to increase with hyperthermic stress [25, 40, 41]. Increased expression of HSP 72 is induced by heat, transient hypoxia, and noise stimulation [40, 42–44]. When mice were treated with heat stress (whole body), protection against noise trauma was demonstrated and an increase in mRNA levels of HSP 70 was found in the cochlea [25].

This heat-induced expression of HSP has been demonstrated to attenuate NIHL in rodents [25]. Quantitative PCR experiments showed a 100- to 200-fold increase of HSP mRNA in the cochlea of mice compared to control mice. The importance of HSP 70 has also been demonstrated in another study using rats. Altschuler et al. [44] showed that rats with high HSP expression (induced by noise) were exposed for a second time (when HSP levels were high), they showed a greater recovery from noise trauma compared to a group without high levels of HSP. When rats were exposed to a second noise exposure, at a time when HPS was not high, no enhanced recovery was found. These results are consistent with the idea that upregulation of heat shock proteins protects the ear from acoustic injury. The exact function of HSPs in the auditory system is unclear, but it may play an important role in protection of the inner ear from further damage and might thus also lie behind the conditioning phenomenon.

Neurotrophic Factors

Application of glial-cell-line-derived neurotrophic factor (GDNF) [45] prior to noise overstimulation has been shown to provide significant functional and cellular protection from acoustic trauma. The recent demonstration that GDNF expression in the cochlea increases following noise exposure [46] supports an endogenous protective role for neurotrophic factors. With the range of neurotrophic factors expressed in the inner ear, it is reasonable to expect that other factors may also undergo changes in expression with stress, thereby protecting the inner ear from permanent damage. However, while the receptor for GDNF has not been localized in the cochlea the underlying mechanism for this protection remains unknown. Neurotrophins may be candidates for protection from NIHL. While neurotrophic factor and antioxidant treatments are typically considered independently, there is evidence suggesting that the two classes of agents may in fact act by similar mechanisms. In some systems, neurotrophic factor deprivation results in cell death due to increased nitric oxide synthase activity [47]. Brain-derived neurotrophic factor (BDNF) has been demonstrated to alter the expression and/or activity of endogenous free radical scavengers, thereby protecting auditory neurons challenged with cisplatin [48].

Amelioration of the effects of oxidative stress is not the only mechanism whereby neurotrophic factors can alter cell survival. Neurotrophic factors can cause changes in the expression of factors driving or preventing programmed cell death [49, 50]. BDNF is believed to alter NMDA-mediated calcium influx and/or downstream signaling [51] and both BDNF and NT-3 can upregulate the expression of calcium-binding proteins [52], thereby altering calcium homeostasis.

The relative contributions of these various protective pathways from noise-induced damage are unclear, at present. Figure 1 illustrates a few of the possible

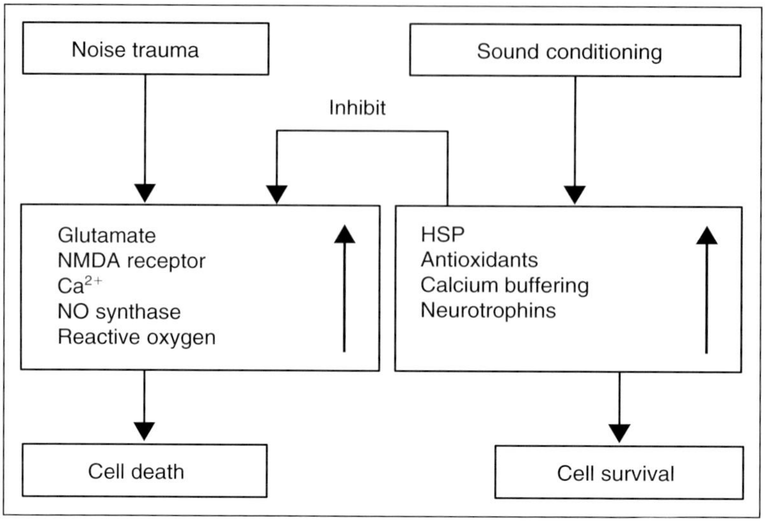

Fig. 1. A few of the biochemical changes that may occur during noise trauma, and how sound conditioning provides protection.

biochemical changes that may occur during noise trauma and shows how sound conditioning may inhibit those damaging factors.

All these events can result in damage to tissues, proteins, lipids, and DNA, partly via membrane lipid peroxidation. Depending upon the severity of the damage, the hair cells may die (necrosis or apoptosis) or survive with varied functional activity. The upregulation of cochlear antioxidant enzymes would be one means of causing a localized protection for all cochlear structures, including the outer hair cells and the efferent nerve endings by sound conditioning. However, a recent study showed that unilateral protection from acoustic trauma was afforded after unilateral sound conditioning [53]. These findings suggest that an overall stress-related mechanism underlying sound conditioning is not the complete cause for protection.

Conclusion

Sound conditioning is one means of protecting against noise trauma, stress, and hereditary hearing loss. Proposed mechanisms include the upregulation of endogenous antioxidants, the number of NMDA receptors, calcium-buffering systems, HSP, and neurotrophic factors. Future experiments are needed to elucidate the mechanisms underlying this phenomenon so that therapeutic interventions and perhaps pharmacological strategies can be developed. It is conveivable that the sound conditioning will benefit human subjects and provide a treatment for NIHL.

References

1 Miller JD, Watson CS, Covell WP: Deafening effects of noise on the cat. Acta Otolaryngol Suppl 1963;176:1–91.

2 Canlon B, Borg E, Flock Å: Protection against noise trauma by pre-exposure to a low level acoustic stimulus. Hear Res 1998;34:197–200.

3 Ryan AF, Bennett TM, Woolf NK, Axelsson A: Protection from noise-induced hearing loss by prior exposure to a nontraumatic stimulus: Role of the middle ear muscles. Hear Res 1994;72: 23–28.

4 Boettcher FA, Schmiedt RA: Distortion-product otoacoustic emissions in Mongolian gerbils with resistance to noise-induced hearing loss. J Acoust Soc Am 1995;98:3215–3222.

5 Dagli S, Canlon B: The effect of repeated daily noise exposure on sound conditioned and unconditioned guinea pigs. Hear Res 1997;104:39–46.

6 Kujawa SG, Liberman MC: Conditioning-related protection from acoustic injury: Effects of chronic deafferentation and sham surgery. J Neurophysiol 1997;78:3095–3106.

7 Canlon B, Borg E, Löfstrand P: Physiological and morphological aspects to low level acoustic stimulation; in Dancer AL, Henderson D, Salvi RJ, Hamernik RP (eds): The Effects of Noise on the Auditory System. St Louis, Mosby-Year Book Inc, 1992.

8 Pukkila M, Zhai S, Virkkala J, Pirovola U, Ylikoski J: The 'toughening' phenomenon in rat's auditory organ. Acta Otolaryngol 1997;529:59–62.

9 White DR, Boettcher FA, Miles LR, Gratton MA: Effectiveness of intermittent and continuous acoustic stimulation in preventing noise-induced hearing and hair cell loss. J Acoust Soc Am 1998; 103:1566–1572.

10 Clark WW, Boettcher FA, Bohne BA: Effect of periodic rest on hearing loss and cochlear damage following exposure to noise. J Acoust Soc Am 1987;82:1253–1264.

11 Sinex DG, Clark WW, Bohne BA: Effect of periodic rest on physiological measures of auditory sensitivity following exposure to noise. J Acoust Soc Am 1987;82:1265–1273.

12 Campo P, Subramaniam M, Henderson D: The effect of 'conditioning' exposures on hearing loss from traumatic exposure. Hear Res 1991;55:195–200.

13 Franklin DJ, Lonsbury-Martin BL, Stagner BB, Martin GK: Altered susceptibility of 2f1–f2 acoustic-distortion products to the effects of repeated noise exposure in rabbits. Hear Res 1991;53: 185–208.

14 Boettcher FA, Spongr VP, Salvi RJ: Physiological and histological changes associated with the reduction in threshold shift during interrupted noise exposure. Hear Res 1992;62:217–236.

15 Boettcher FA: Auditory brain-stem response correlates of resistance to noise-induced hearing loss in the Mongolian gerbil. J Acoust Soc Am 1993;94:3207–3214.

16 Subramaniam M, Campo P, Henderson D: The effect of exposure level on the development of progressive resistance to noise. Hear Res 1991;52:181–188.

17 Subramaniam M, Henderson D, Campo P, Spongr V: The effect of 'conditioning' on hearing loss from a high frequency traumatic exposure. Hear Res 1992;58:57–62.

18 Miyakita T, Hellström PA, Frimanson E, Axelsson A: Effect of low level acoustic stimulation on temporary threshold shift in young humans. Hear Res 1992;60:149–155.

19 Henselman LW, Henderson D, Subramaniam M, Sallustio V: The effect of 'conditioning' exposures on hearing loss from impulse noise. Hear Res 1994;78:1–10.

20 Henderson D, Subramaniam M, Papazian M, Spongr VP: The role of middle ear muscles in the development of resistance to noise induced hearing loss. Hear Res 1994;74:22–28.

21 McFadden SL, Henderson D, Shen YH: Low-frequency 'conditioning' provides long-term protection from noise-induced threshold shifts in chinchillas. Hear Res 1997;103:142–150.

22 Fowler T, Canlon B, Dolan D, Miller J: The effects of noise trauma following training exposures in the mouse. Hear Res 1995;88:1–13.
23 Yoshida N, Liberman MC: Sound conditioning reduces noise-induced permanent threshold shift in mice. Hear Res 2000;148:213–219.
24 Willott JF, Turner JG, Sundin VS: Effects of exposure to an augmented acoustic environment on auditory function in mice: Roles of hearing loss and age during treatment. Hear Res 2000;142: 79–88.
25 Yoshida N, Kristiansen A, Liberman MC: Heat stress and protection from permanent acoustic injury in mice. J Neurosci 1999;19:10116–10124.
26 Yamane H, Nakai Y, Takayama M, Iguchi H, Nakagawa T, Kojima A: Appearance of free radicals in the guinea pig inner ear after noise-induced acoustic trauma. Eur Arch Otorhinolaryngol 1995; 252:504–508.
27 Ohlemiller KK, McFadden SL, Ding DL, Flood DG, Reaume AG, Hoffman EK, Scott RW, Wright JS, Putcha GV, Salvi RJ: Targeted deletion of the cytosolic Cu/Zn-superoxide dismutase gene (Sod1) increases susceptibility to noise-induced hearing loss. Audiol Neurootol 1999;4: 237–246.
28 Yamasoba T, Harris C, Shoji F, Lee RJ, Nuttall AL, Miller JM: Influence of intense sound exposure on glutathione synthesis in the cochlea. Brain Res 1998;804:72–78.
29 Yamasoba T, Nuttall AL, Harris C, Raphael Y, Miller JM: Role of glutathione in protection against noise-induced hearing loss. Brain Res 1998;784:82–90.
30 Yamasoba T, Schacht J, Shoji F, Miller JM: Attenuation of cochlear damage from noise trauma by an iron chelator, a free radical scavenger and glial cell line-derived neurotrophic factor in vivo. Brain Res 1999;815:317–325.
31 Hu BH, Zheng XY, McFadden SL, Kopke RD, Henderson D: *R*-phenylisopropyladenosine attenuates noise-induced hearing loss in the chinchilla. Hear Res 1997;113:198–206.
32 Seidman MD, Shivapuja BG, Quirk WS: The protective effects of allopurinol and superoxide dismutase on noise-induced cochlear damage. Otolaryngol Head Neck Surg 1993;109:1052–1056.
33 Jacono AA, Hu B, Kopke R, Henderson D, Van De Water T, Steinman HM: Changes in cochlear antioxidant enzyme activity after sound conditioning and noise exposure in the chinchilla. Hear Res 1998;117:31–38.
34 Lautermann J, Crann SA, McLaren J, Schacht J: Glutathione-dependent antioxidant systems in the mammalian inner ear: Effects of aging, ototoxicity, and noise. Hear Res 1997;114:75–82.
35 Ernfors P, Canlon B: Aminoglycoside excitement silences hearing. Nat Med 1996;2:1313–1314.
36 Hu HB, Henderson D: Changes in F-actin labeling in the outer hair cell and the Deiters cell in the chinchilla cochlea following noise exposure. Hear Res 1997;110:209–218.
37 Parsell DA, Kowal AS, Singer MA, Lindquist S: Protein disaggregation mediated by heat-shock protein Hsp104. Nature 1994;372:475–478.
38 Fawcett TW, Sylvester SL, Sarge KD, Morimoto RI, Holbrook NJ: Effects of neurohormonal stress and aging on the activation of mammalian heat shock factor 1. J Biol Chem 1994;269: 32272–32278.
39 Neely JG, Thompson AM, Gower DJ: Detection and localization of heat shock protein 70 in the normal guinea pig cochlea. Hear Res 1991;52:403–406.
40 Dechesne CJ, Kim HN, Nowak TS Jr, Wenthold RJ: Expression of heat shock protein, HSP72, in the guinea pig and rat cochlea after hyperthermia: Immunochemical and in situ hybridization analysis. Hear Res 1992;59:195–204.
41 Thompson AM, Neely JG: Induction of heat shock protein in interdental cells by hyperthermia. Otolaryngol Head Neck Surg 1992;107:769–774.
42 Myers MW, Quirk WS, Rizk SS, Miller JM, Altschuler RA: Expression of the major mammalian stress protein in the rat cochlea following transient ischemia. Laryngoscope 1992;102:981–987.
43 Lim HH, Jenkins OH, Myers MW, Miller JM, Altschuler RA: Detection of HSP 72 synthesis after acoustic overstimulation in rat cochlea. Hear Res 1993;69:146–150.
44 Altschuler RA, Lim HH, Ditto J, Dolan D, Raphael Y: Protective mechanisms in the cochlea: Heat shock proteins; in Salvi RJ, Henderson D, Fiorino F, Colletti A (eds): Auditory System Plasticity and Regeneration. Stuttgart, Thieme, 1996, pp 202–212.

45 Keithley EM, Ma CL, Ryan AF, Louis JC, Magal E: GDNF protects the cochlea against noise damage. Neuroreport 1998;9:2183–2187.
46 Nam YJ, Stover T, Hartman SS, Altschuler RA: Upregulation of glial cell line-derived neurotrophic factor (GDNF) in the rat cochlea following noise. Hear Res 2000;146:1–6.
47 Estevez AG, Spear N, Manuel SM, Barbeito L, Radi R, Beckman JS: Role of endogenous nitric oxide and peroxynitrite formation in the survival and death of motor neurons in culture. Prog Brain Res 1998;118:269–280.
48 Gabaizadeh R, Staecker H, Liu W, Van De Water TR: BDNF protection of auditory neurons from cisplatin involves changes in intracellular levels of both reactive oxygen species and glutathione. Brain Res Mol Brain Res 1997;50:71–78.
49 Mattson MP, Culmsee C, Yu ZF: Apoptotic and antiapoptotic mechanisms in stroke. Cell Tissue Res 2000;301:173–187.
50 Rong P, Bennie AM, Epa WR, Barrett GL: Nerve growth factor determines survival and death of PC12 cells by regulation of the bcl-x, bax, and caspase-3 genes. J Neurochem 1999;72: 2294–2300.
51 Brandoli C, Sanna A, De Bernardi MA, Follesa P, Brooker G, Mocchetti I: Brain-derived neurotrophic factor and basic fibroblast growth factor downregulate NMDA receptor function in cerebellar granule cells. J Neurosci 1998;18:7953–7961.
52 Nakao N, Kokaia Z, Odin P, Lindvall O: Protective effects of BDNF and NT-3 but not PDGF against hypoglycemic injury to cultured striatal neurons. Exp Neurol 1995;131:1–10.
53 Yamasoba T, Dolan DF, Miller JM: Acquired resistance to acoustic trauma by sound conditioning is primarily mediated by changes restricted to the cochlea, not by systemic responses. Hear Res 1999;127:31–40.

Xianzhi Niu, Department of Physiology and Pharmacology,
Karolinska Institutet, S–17177 Stockholm (Sweden)
Tel. +46 8 728 7248, Fax +46 8 327026, E-Mail niuxianzhi@hotmail.com

Felix D, Oestreicher E (eds): Rational Pharmacotherapy of the Inner Ear.
Adv Otorhinolaryngol. Basel, Karger, 2002, vol 59, pp 106–111

Effects of Aging on C57BL/6J Mice: An Electrophysiological and Morphological Study

María Visitación Bartolomé, Eduardo del Castillo, Lorena Maestre López, Francisco Carricondo, Joaquin Poch-Broto, Pablo Gil-Loyzaga

Center of Cell Culture (CAI-UCM) and Department of Surgery II (ORL). Faculty of Medicine, University Complutense, Madrid, Spain

Abstract

Presbycusis is a progressive hearing loss associated with aging that manifests as deafness linked to cochlear morphological degeneration. The effects of aging on the auditory system were studied in C57BL/6J mice using electrophysiological (brainstem auditory evoked potentials; BAEP) and morphological techniques. Cochleae of animals aged 1, 6, 9, 12, 15, 18, 21, or 24 months old were used for that purpose. The BAEP showed a progressive increase in latency and a reduction in amplitude. Morphological studies demonstrated total degeneration of the organ of Corti, which was replaced by a single epithelial layer. An affinity histochemistry study demonstrated minor modifications of glycoconjugates in the organ of Corti during the aging process.

Presbycusis is a well-known progressive, bilateral and symmetric deafness which frequently appears in aged mammals. Four main types of presbycusis were described: sensory, neural, metabolic and mechanical [1, 2]. Presbycusis is characterized by elevation of hearing thresholds, starting at high frequencies and progressing to lower frequencies [3, 4]. Physiological findings are highly correlated with histopathology, which includes progressive sensorineuronal degeneration, including important changes in sensory cells, pillar cells, stria vascularis, spiral ligament and basilar membrane [5–7], and the loss of spiral ganglion neurons [1, 8]. However, the age-related changes in the biochemical composition of the tectorial membrane (TM) have seldom been investigated.

A large study of the TM has allowed to directly correlate its biochemical composition to its physiology during the auditory process [9]. Lectins are proteins of nonimmune origin that specifically bind carbohydrates [10] and are widely used to analyze the TM composition of cochleae in both normal and hypothyroid rats (during development or at adulthood) [9, 11–14], and in human cochleae [15].

A preliminary study of the effects of aging on the glycoconjugate composition of the TM could provide information about molecular changes in presbycusis. For that purpose, the C57BL/6J mouse strain, which exhibits premature aging effects in the auditory system, could be an appropriate experimental model [3, 16]. In the present study, the brainstem auditory evoked potential (BAEP) recordings were correlated to the histopathological degeneration of the sensory epithelium of the organ of Corti and the presence of different glycoconjugates in the C57BL/6J mouse strain cochlea.

Materials and Methods

A total of 48 C57BL/6J adult mice were divided into eight age groups: 1, 6, 9, 12, 15, 18, 21, and 24 months.

Electrophysiological Study

The electrophysiological procedure was carried out under deep anesthesia, in an anechoic chamber. BAEP recordings were obtained using clicks with stimulus intensities ranging between 100 and 30 dB (SPL). Latency and amplitude of the five waves were measured. Statistics were based on a one-way Anova with a Bonferroni post hoc test. Thereafter, the cochleae were quickly removed and submitted to a morphological study.

Histological and Affinohistochemical Studies

In the present paper, a total of 12 cochleae of C57BL/6J mice of 1 or 24 months of age were used. The cochleae were rapidly removed and fixed in 2% acetic acid in 70% ethanol for 72 h and decalcified by EDTA (4% in PBS). Paraffin sections (7 μm thick), were obtained from the mid-modiolar plane. One section was counterstained with anilins.

Some sections were used for affinity histochemical studies with fluorescein-labeled lectins as previously described [11]. The selected lectins were: DBA (*Dolichos biflorus*, horse gram), SBA (*Glycine max*, soybean), PNA (*Arachis hypogea*, peanut), UEA-1 (*Ulex europeus*, gorse seed), RCA-1 (*Ricinus communis*, castor bean), WGA (*Triticum vulgare*, wheat germ) (Vector Laboratories).

Results

Electrophysiological Results

The BAEP recordings showed a significant decrease in the amplitude and an increase in the latency in mice aged 15, 18, 21 and 24 months, with respect to the recordings on 1, 6, 9, 12 months old mice (data not shown).

Histological and Affinohistochemical Studies

The organ of Corti of 1-month-old mice exhibited a normal shape, including pillar, Deiters's and sensory cells (fig. 1A). In contrast, cochleae of 24-month-old mice exhibited a degenerated basal coil, showing a single epithelial layer which covers the basilar membrane (fig. 1B).

Lectin studies on 1-month-old mice cochleae showed a positive DBA lectin binding in the TM limbal zone (fig. 1C); a similar, but very weak, reaction was observed in cochleae of 24-month-old mice (fig. 1D). When SBA lectin was used, only the TM limbal and middle zones were positive in 1-month-old mice (fig. 1E); there was no reaction in 24-month-old mice (fig. 1F). A very small patch appeared in the innermost part of the limbal TM (fig. 1F). In addition, both RCA-1 (fig. 1G) and WGA (fig. 1H) lectins exhibited a very strong fluorescent labeling up to 24 months (figs. 1G, 1H). UEA-1 and PNA lectins did not produce any significant labeling of the organ of Corti.

Discussion

In the present study, the first electrophysiological alterations of auditory function were observed in 15-month-old mice. A progressive increase in latency and decrease in amplitude of the BAEP waves were detected during aging. These results fit well with previous studies and highly correlate to morphological findings. In fact, a progressive degeneration of the organ of Corti, including supporting and sensory cells, which first affects the outer hair cells and then the

Fig. 1. Mid modiolar paraffin sections of the organ of Corti at the basal coil of a C57BL/6J mouse cochlea. At 1 month, the sensory epithelium, the tunnel of Corti and supporting cells (***A***) exhibit their habitual shape. At 24 months, a single epithelial layer constitutes a degenerative scar (***B***). The DBA lectin reaction is visible in the TM limbal zone in the cochlea of a 1-month-old mouse (***C***), reduced in a 24-month-old mouse (***D***, arrow). The SBA lectin labeling appears in the TM limbal and middle zones of a 1-month-old mouse (***E***), but is totally absent at 24 month (***F***, arrow). A strong RCA-1 (***G***) and WGA (***H***) lectin label is present in the whole TM in a 24-month-old mouse. I = Inner hair cell; O = outer hair cell. TM = Tectorial membrane; star = tunnel of Corti. Scale bar: 50 μm.

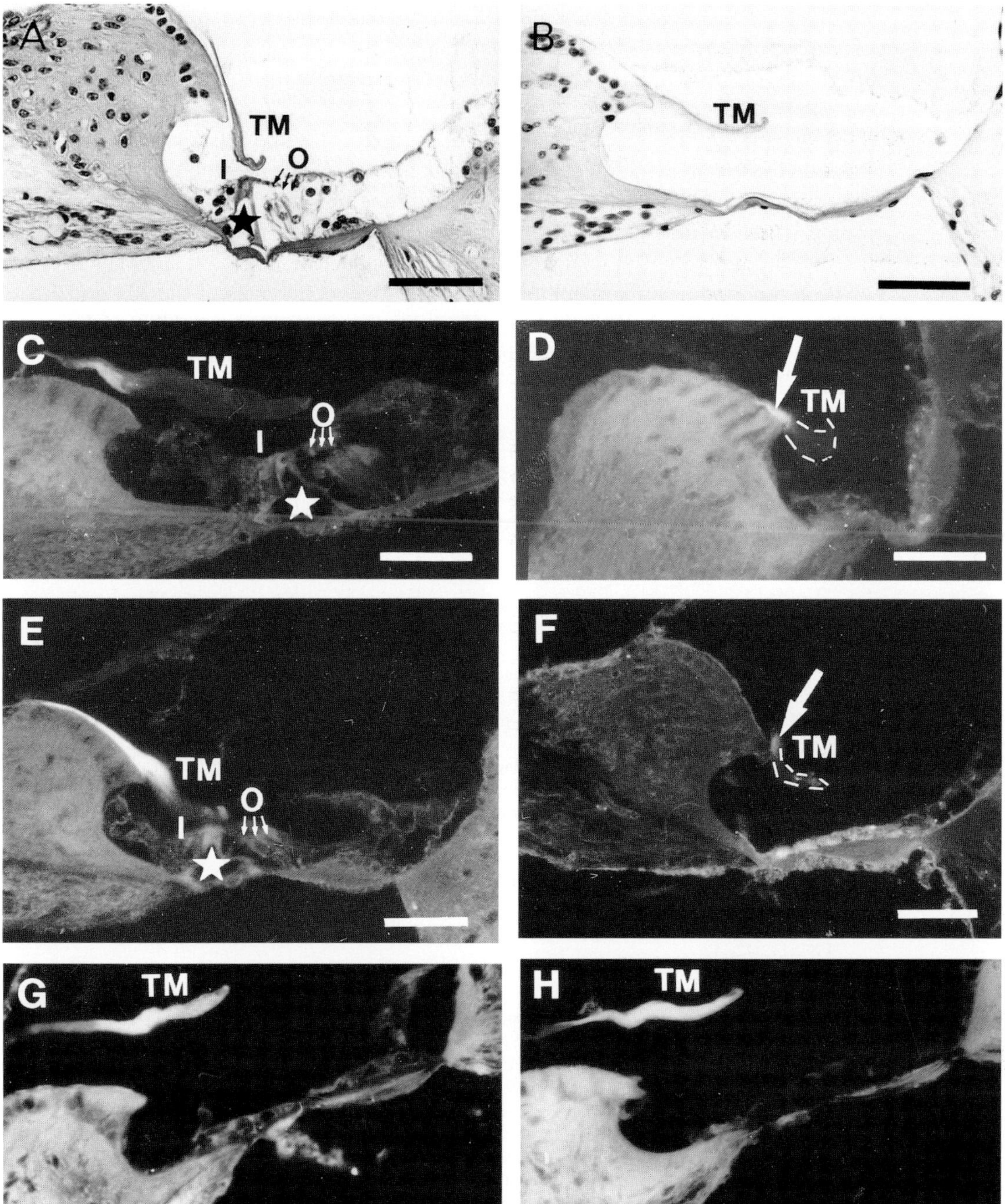
A
TM
I
O
B
TM
C
TM
O
I
D
TM
E
TM
O
I
F
TM
G
TM
H
TM

inner hair cells, started at the base coil [3, 5, 6]. An epithelial scar covering the basilar membrane was clearly detected, suggesting that a progressive transdifferentiation of the sensory epithelium also appears with aging similarly to ototoxicity [17].

The pattern of the reactivity observed for the RCA-1 and WGA lectins corresponded to that in the adult euthyroid animals [9, 11]. The present observation of the reduction of the DBA and SBA lectin labeling could be an interesting finding which in fact could be related to a change in the TM glycoconjugate structure, in particular a significant reduction in galactose levels [9]. This sugar has been reported as a main component of the glycoconjugates of the TM and has been involved in its function [9]. Thus, this kind of observation could be linked to the general progressive degenerative process detected in presbycusis. A detailed study must be carried out to provide additional information.

Acknowledgements

This study has been supported by grants PM99-0054 and FIS 98/0732.

References

1 Johnsson LG, Hawkins JE Jr: Sensory and neural degeneration with aging, as seen in microdissections of the human inner ear. Ann Otol Rhinol Laryngol 1972;81:79–193.
2 Schuknecht HF, Gacek MR: Cochlear pathology in presbycusis. Ann Otol Rhinol Laryngol 1993;102:1–14.
3 Henry KR, Chole RA: Genotypic differences in behavioral, physiological and anatomical expressions of age-related hearing loss in the laboratory mouse. Audiology 1980;19:369–383.
4 Willot JF, Erway LC, Archer JR, Harrison DE: Genetics of age-related hearing loss in mice. II. Strain differences and effects of caloric restriction on cochlear pathology and evoked response. Hear Res 1995;88:143–155.
5 Bhattacharya TK, Dayal VS: Influence of age on hair cell loss in the rabbit cochlea. Hear Res 1989;40:179–184.
6 Ingham NJ, Comis SD, Withington: Hair cell loss in the guinea pig cochlea. Acta Otolaryngol 1999;119:42–47.
7 Ichimiya I, Suzuki M, Mogi G: Age-related changes in the murine cochlear lateral wall. Hear Res 2000;139:116–122.
8 Felix H, Johnsson LG, Gleeson M: Quantitative analysis of cochlear sensory cells and neural elements in man. Acta Otolaryngol Suppl (Stockh) 1990;470:71–79.
9 Gil-Loyzaga P: Histochemistry of glycoconjugates of the auditory receptor-functional implications. Prog Histochem Cytochem 1997;32:1–79.
10 Goldstein IJ, Hughes RC, Monsigny M, Osawa T, Sharon N: What should be called a lectin? Nature 1980;285: 66.
11 Gil-Loyzaga P, Gabrion J, Uziel A: Lectins demonstrate the presence of carbohydrates in the tectorial membrane of mammalian cochlea. Hear Res 1985;20:1–8.
12 Gil-Loyzaga P, Menoyo A, Poch-Broto J, Merchán-Perez A: Effects of the perinatal hypothyroidism in the carbohydrate composition of cochlear tectorial membrane. Hear Res 1990;45: 151–156.

13 Remezal M, Gil-Loyzaga P, Mollicone R, Joubert-Caron R, Oriol R: Histochemical localization of galactoside-binding lectin and its binding-sites in the developing and adult rat cochlea. Dev Brain Res 1993;73:1–6.
14 Khan KM, Sarfaraz N, Salim Z: Lectin binding patterns in nonsensory regions of rat cochlea during postnatal development. J Anat 1999;194:479–504.
15 Ibáñez A, Rodríguez-Ramos A, Bartolomé MV, Valderrama F, Remezal M, Micó C, Dargallo J, Salmean J: Glycoconjugates as components of the tectorial membrane of the human fetal cochlea; in Progress in Human Audiology and Vestibular Histopathology. Amsterdam, Kugler, 1997, pp 153–156.
16 Henry KR, McGinn MD: The mouse as a model of human audition. Audiology 1992;31:181–189.
17 Bartolomé MV, Vago P, Gil-Loyzaga P, Humbert G, Joubert-Caron R, Pujol R, Lenoir M: Sequential changes in anti-GAL-1 staining of the rat organ of Corti following amikacin exposure. Brain Res 1999;822:43–51.

Prof. P. Gil-Lloyzaga, MD, PhD, Universidad Complutense,
Apartado de Correos 60.075, E–28080 Madrid (Spain)
Tel. +34 91 3941 375, Fax +34 91 3941 383, E-Mail loyzaga@eucmax.sim.ucm.es

Felix D, Oestreicher E (eds): Rational Pharmacotherapy of the Inner Ear.
Adv Otorhinolaryngol. Basel, Karger, 2002, vol 59, pp 112–115

Effects of Aging on Cochlear Monoamine Turnover

M. Angeles Vicente-Torres, David Dávila, Encarnación Muñoz, Pablo Gil-Loyzaga

Center of Cell Culture (CAI-UCM) and Department of Surgery II,
Faculty of Medicine, University Complutense, Madrid, Spain

Abstract

The aging of the cochlear dopaminergic system has been analyzed by quantifying the levels of dopamine (DA) and its metabolites (3,4-dihydroxyphenylacetic acid, DOPAC, and homovanillic acid, HVA) in adult rats aged 3, 12 or 24 months. The main results were an increase in DA, DOPAC and HVA basal concentrations in aged females with respect to the adults (3 or 12 months old), while just DA and DOPAC increased in aged males. A higher synthesis of DA in aged animals could support these findings, which could indicate some kind of compensatory mechanism related to presbycusis.

Introduction

Cochlear dopaminergic fibers belonging to the olivocochlear efferent bundle have been involved in the modulation of primary auditory neuron activity [1–6]. Basal levels of DA and its two main metabolites 3,4-dihydroxyphenylacetic acid (DOPAC) and homovanillic acid (HVA) were quantified by high-performance liquid chromatography with electrochemical detection (HPLC-ED) in rat cochleae [1, 5, 7, 8]. Levels of basal DA and its metabolites changed due to acoustic stimulation indicating that cochlear DA turnover is linked to noise [1, 2, 4, 5]. In addition, a protective role for primary auditory neurons in potentially harmful situations, e.g. acoustic trauma, neurotoxicity or hypoxia, has been suggested [3, 5, 9].

Dopaminergic neurotransmission in other nervous system areas suffers several anomalies during the aging process, which were manifested as changes in DA, DOPAC and HVA levels [10, 11]. However, discrepancies have been reported depending on the gender and strain of the animals, the nervous area analyzed and the experimental protocols.

The aim of the present study was to analyze the effect of aging on the dopaminergic system projecting to the cochlea.

Material and Methods

Thirty Long-Evans rats were divided into three groups aged 3, 12 or 24 months (5 males and 5 females were analyzed at each age). The animals were kept in silence (less than 20 dB SPL) for 1 h, under general anesthesia with urethane, to keep basal acoustic conditions. Then, still under general anesthesia, the cochleae were quickly removed and individually collected in chilled 0.2 *N* perchloric acid. They were crushed, sonicated and centrifuged. Supernatants were filtered and frozen at −80 °C and stored until analysis (not more than 3 days).

HPLC-ED was used to quantify the concentrations of DA, DOPAC and HVA in cochlear supernatants, as previously described [8, 12]. Statistical analysis was carried out by two-way Anova (age × gender) followed by one-way Anova when interaction was detected between main factors. Tukey's test was used for post hoc analysis.

Results

The cochlear DA concentration was higher in the oldest animals (24 months) than in 3- or 12-month-old ones (table 1). The concentration of DOPAC was also higher in the cochleae of aged animals than in those of younger ones (table 2). No significant gender-related difference was observed in cochlear DA and DOPAC concentrations.

The cochlear concentration of HVA was independent of the gender in young and adult animals (table 3). However, a gender effect was observed when the HVA concentration was analyzed in aged animals. Aged females showed a higher HVA concentration than younger ones, while males did not show any age-related modification of their cochlear HVA concentration (table 3).

Discussion

The cochlear dopaminergic system has shown some particular changes in aged (24 months) rats. In contrast, no differences were observed between young (3 months) and adult (12 months) animals. Cochleae of aged animals (24 months) showed an increase in the concentration of DA and its metabolites (DOPAC and HVA in females and just DOPAC in males).

Table 1. Cochlear DA concentration (pg/cochlea)

	3 months old	12 months old	24 months old
Females	62.46 ± 5.20	66.14 ± 5.64	103.22 ± 19.4*
Males	61.90 ± 2.76	65.61 ± 6.11	91.44 ± 8.65*

*p < 0.001 with respect to younger animals. Means ± SEM.

Table 2. Cochlear DOPAC concentration (pg/cochlea)

	3 months old	12 months old	24 months old
Females	36.27 ± 6.69	39.58 ± 4.37	64.13 ± 4.41*
Males	36.44 ± 3.42	39.01 ± 6.12	52.94 ± 7.39*

*p < 0.001 with respect to younger animals. Means ± SEM.

Table 3. Cochlear HVA concentration (pg/cochlea)

	3 months old	12 months old	24 months old
Females	51.19 ± 2.69	53.90 ± 5.80	75.87 ± 9.53*
Males	53.28 ± 6.60	57.56 ± 9.16	47.62 ± 8.94

*p < 0.05 with respect to younger females. Means ± SEM.

These changes in the cochlear DA system could indicate an important modification of synthesis activity of dopaminergic neurons of the superior olivary complex. This increase could constitute a compensatory mechanism to protect the afferent type I neurons against age-induced damage. In this way, the nigrostriatal system exhibited an increase in DA synthesis as a result of a compensatory effect against the age-related loss of neurons in the substantia nigra [13, 14]. Gender-dependent differences could indicate a different compensatory power or an unlike primary damage in the cochlear afferent fibers between females and males.

The present findings could be relevant in order to clarify the mechanisms involved in neural presbycusis and to design a pharmacological approach to contribute to the prevention of this disorder.

Acknowledgements

The present work has been supported by grants PM 990054 and FIS 98/0732.

References

1 Gil-Loyzaga P, Fernández-Mateos P, Vicente-Torres MA, Remezal M, Cousillas H, Arce A, Esquifino A: Effects of noise stimulation on cochlear dopamine metabolism. Brain Res 1993; 623:177–180.

2 Gil-Loyzaga P, Vicente-Torres MA, Fernández-Mateos P, Arce A, Esquifino A: Piribedil affects dopamine turnover in cochleas stimulated by white noise. Hear Res 1994;79:178–182.

3 Pujol R, Puel JL, Gervais d'Aldin C, Eybalin M: Pathophysiology of the glutamatergic synapses in the cochlea. Acta Otolaryngol (Stockh) 1993;113:330–334.

4 Vicente-Torres MA, Gil-Loyzaga P, Fernández P, Arce A, Esquifino A: Piribedil could modify dopamine turnover in cochleas under noise stimulation. J Otorhinolaryngol Relat Spec 1993;55: 322–324.

5 Gil-Loyzaga P: Neurotransmitters of the olivocochlear lateral efferent system: With an emphasis on dopamine. Acta Otolaryngol Suppl (Stockh) 1995;115:222–226.

6 Oestreicher E, Arnold W, Ehrenberger K, Felix D: Dopamine regulates the glutamatergic inner hair cell activity in guinea pigs. Hear Res 1997;107:46–52.

7 Gil-Loyzaga P, Parés Herbute N: HPLC detection of dopamine and noradrenaline in the cochlea of adult and developing rats. Dev Brain Res 1989;48:157–160.

8 Gil-Loyzaga P, Vicente-Torres MA, Arce A, Cardinali DP, Esquifino A: Effect of superior cervical ganglionectomy on catecholamine concentration in rat cochlea. Brain Res 1998;779:53–57.

9 Eybalin M, Safieddine S, D'Aldin CG, Pujol R, Puel JL: Molecular mechanisms involving neurotransmitters after excitotoxicity at the inner hair cell-auditory nerve synapse; in Flock A, Ottoson D, Ulfendahl M (eds): Active Hearing. Amsterdam, Pergamon Press, 1995, pp 73–85.

10 Míguez JM, Aldegunde M, Paz-Valiñas L, Recio J, Sánchez-Barceló E: Selective changes in the contents of noradrenaline, dopamine and serotonin in rat brain areas during aging. J Neural Transm 1999;106:1089–1098.

11 van Luijtelaar MGPA, Tonnaer JADM, Steinbusch HWM: Aging of the serotonergic system in the rat forebrain: An immunocytochemical and neurochemical study. Neurobiol Aging 1992;13: 201–215.

12 Gil-Loyzaga P, Vicente-Torres MA, García-Bonacho M, Esquifino A: Presence of catecholamines and serotonin in the rat vestibule. Brain Res 1997;746:265–268.

13 Greenwood CE, Tatton WG, Seniuk NA, Biddle FG: Increased dopamine synthesis in aging substantia nigra neurons. Neurobiol Aging 1991;12:557–565.

14 Tatton WG, Greenwood CE, Salo PT, Seniuk NA: Transmitter synthesis increases in substantia nigra neurons of the aged mouse. Neurosci Lett 1991;131:179–182.

P. Gil-Loyzaga, Apartado de correos 60075, E–28080 Madrid (Spain)
Tel. +34 91 394 13 75, Fax +34 91 394 13 83, E-Mail loyzaga@eucmax.sim.ucm.es

Felix D, Oestreicher E (eds): Rational Pharmacotherapy of the Inner Ear.
Adv Otorhinolaryngol. Basel, Karger, 2002, vol 59, pp 116–123

Degeneration Pattern of Human First-Order Cochlear Neurons

Heidi Felix[a], *Anita Pollak*[a], *Michael Gleeson*[b], *Lars-Göran Johnsson*[c]

[a] ENT Department, University Hospital, Zürich, Switzerland;
[b] Guy's Hospital, London, UK;
[c] University of Helsinki, Finland

Abstract

In the present study, quantitative analysis of the cochlear neurons in the osseous spiral lamina, the modiolus and the internal auditory canal of the same cochlea was performed. Forty-five temporal bones were obtained from 25 patients and prepared by means of microdissection. Ten patients had age-related normal hearing (ARNH) assuming that the 5 children without audiogram had normal hearing. Fifteen patients had sensorineural hearing loss due to various causes.

The present study has shown that in young individuals the numbers of cochlear neurons are almost identical at all 3 sites. In patients over 60 years with ARNH, the loss of peripheral nerve processes is always severer than the loss of central nerve processes. This finding suggests that the central processes degenerate at a much slower rate or not at all.

Furthermore, 4 different peripheral degeneration patterns were described. The factors responsible for the different degeneration behaviors are still not understood and need further investigation.

The pattern and sequence of degeneration of the human first-order cochlear neurons are still not well understood. The time course of retrograde degeneration of the cochlear neurons may differ widely, both by etiology and by species. Several authors [1–3] have shown that in the guinea pig and cat degeneration of spiral ganglion cells (SGCs) may occur over months to years following ototoxic drug administration. On the other hand, few authors reported a rapid degeneration of SGCs in the rat [4] and cat [5] as a result of various lesions to the organ of Corti. For humans, several authors described that the number of remaining

SGCs far exceeded the number of remaining dendritic fibers suggesting that SGCs may survive the loss of their dendritic processes for years [6–10]. However, in a series of papers, Spoendlin and Schrott [11–13] suggested that the loss of peripheral processes is accompanied by the loss of SGCs and their central processes. Most of the quantitative studies reported so far used either the SGCs or nerve fibers alone. The aim of this study was to perform quantitative and qualitative analysis of the human cochlear neurons at the osseous spiral lamina (OSL), spiral ganglion and at the internal auditory canal (IAC) within the same cochlea.

Material and Methods

In the present study, quantitative assessment of the neuronal structures in 42 human cochleas of 25 patients was performed. All patients had pure-tone audiometric tests within 12 months prior to their death with the exception of 7 patients. Ten patients had age-related normal hearing (ARNH) assuming that the 5 children and the 2 adults without audiogram had normal hearing. Fifteen patients including 1 without audiogram had sensorineural hearing loss (SNHL) of different degree, cause or type.

The temporal bones were obtained at autopsy within 2–20 h postmortem and were immediately fixed by perilymphatic perfusion through the round and oval windows using 3% glutaraldehyde in 0.1 *M* phosphate buffer at pH 7.3. Afterwards the temporal bones were kept in the same fixation solution until the microdissection was carried out. Prior to the preparation of the labyrinth, the facial and cochleovestibular nerves were dissected from the IAC and further processed for light and electron microscopy [7].

The cochleas were subjected to a complete microdissection after postfixation with 1% phosphate-buffered osmium tetroxide. The method has been described in detail elsewhere [14, 15]. The remaining modioli were decalcified and embedded in Epon and serially cut into 2-μm-thick sections.

The total counts of myelinated nerve fibers of the cochlear nerve in the IAC were determined in light-microscopic photomontages of semithin cross-sections of the cochleovestibular nerve at the central end of Scarpa's ganglion. The myelinated nerve fibers in the OSL were counted in tangential, semithin sections by means of an image analysis system (ASBA II, Wild-Leitz). The SGCs were counted in serial sections of the modiolus.

The total counts of the cochlear neurons at the 3 sites were compared within the same cochlea and between cochleas from patients with ARNH and SNHL.

Results

Nerve Fiber Counts in the IAC and OSL

Some of the nerve fiber counts in the OSL and IAC have been reported earlier [15]. In the meantime, the counts from 8 cochleas of 5 young children ranging in age from 1 day to 2 years could be added to the existing results. These

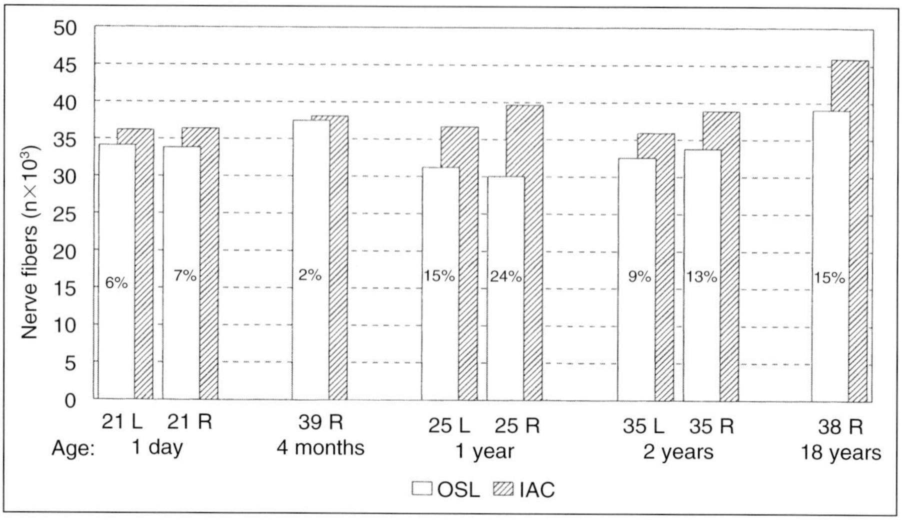

Fig. 1. Diagram showing the differences in percentage of the number of nerve fibers in the IAC and the OSL in the same cochlea of young individuals (1 day to 18 years) with ARNH. L = Left; R = right.

additional data appear to be important for the understanding of the results obtained from older patients.

Age-Related Normal Hearing. In all cases, the number of nerve fibers was higher in the IAC than in the OSL. In the group of children, the number of nerve fibers varied between 35,900 and 45,900 in the IAC, and between 30,000 and 39,000 in the OSL. The differences varied between 2 and 24%. In 4 out 8 cochleas, a difference larger than 10% was present. The smallest differences were found in the ears of babies aged 1 day and 4 months (fig. 1).

The counts of nerve fibers were higher at both levels compared to the group of patients over 60 years.

In the group of patients over 60 years, the number of nerve fibers decreased with age at both levels, but the degeneration appeared slower in the IAC, which was recognizable by the increased differences between the two levels. The differences varied between 14 and 34%. The number of fibers ranged from 25,400 to 32,700 in the IAC and from 18,300 to 27,500 in the OSL (fig. 2).

Sensorineural Hearing Loss. The group of 15 patients with SNHL of different causes and types included 25 cochleas. In all cochleas, the number of nerve fibers was lower in the OSL than in the IAC. In the latter the values varied between 21,600 and 40,000 and were only slightly lower than in the ARNH group. However, the number of fibers was much lower in the OSL ranging

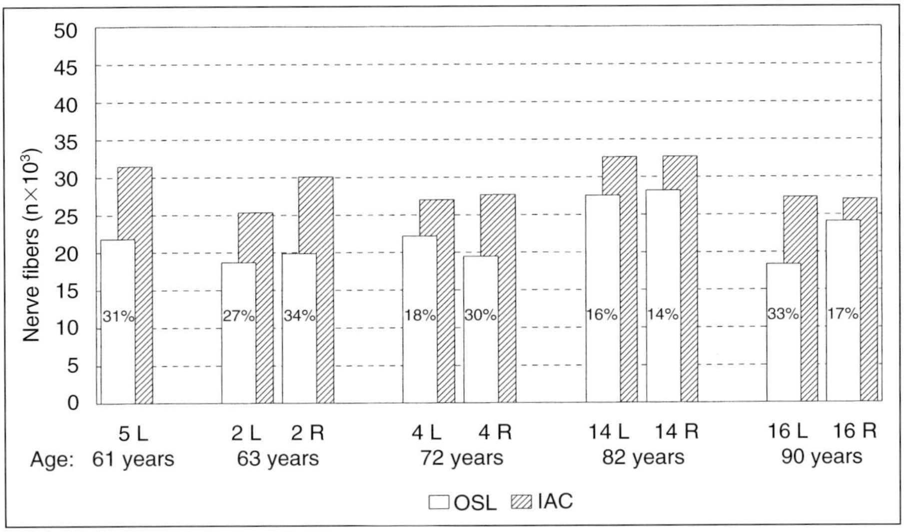

Fig. 2. Diagram showing the differences in percentage of the number of nerve fibers in the IAC and the OSL in the same cochlea of patients over 60 years (61–90) with ARNH. L = Left; R = right.

between 9,422 and 23,968 compared to the values in the ARNH group. In 17 cochleas, the degeneration of dendrites was much greater in relation to the degeneration of axons leading to a marked difference between 36 and 74%. From the remaining 8 cochleas, 3 had counts which were similar to the values in the ARNH group. On the other hand, the 5 cochleas had a great loss of nerve fibers in the OSL as well as in the IAC, and for that reason only a small difference between the two sites was present (fig. 3).

Counts of Neurons in the IAC, Modiolus and OSL

Age-Related Normal Hearing. The quantitative evaluation of 6 cochleas from 6 patients with ARNH at all 3 sites revealed 2 patterns. In all cases, the number of SGCs was similar or slightly higher than the number of nerve fibers in the OSL. In 2 very young children (1 day and 4 months), the counts in all 3 levels were very similar and were defined as pattern 1. In the patients over 60 years with ARNH, the counts of the dendrites and SGCs were comparable, but the number of fibers in the IAC far exceeded the number of the other 2 sites and was characterized as pattern 2 (fig. 4).

Sensorineural Hearing Loss. In 11 cochleas from 10 patients, the quantitative analysis was performed at all 3 levels. In this group of patients, 2 additional degeneration patterns could be observed. In 1 cochlea, pattern 2 was found which

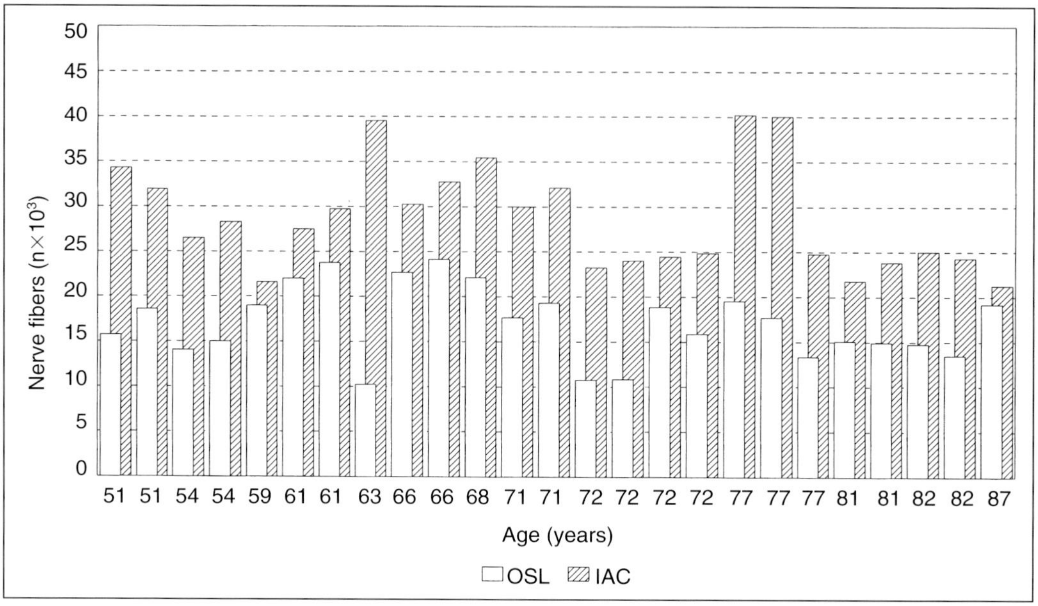

Fig. 3. Diagram showing total counts of nerve fibers in the IAC and OSL in the same cochlea of patients (51–87 years) with SNHL of various causes.

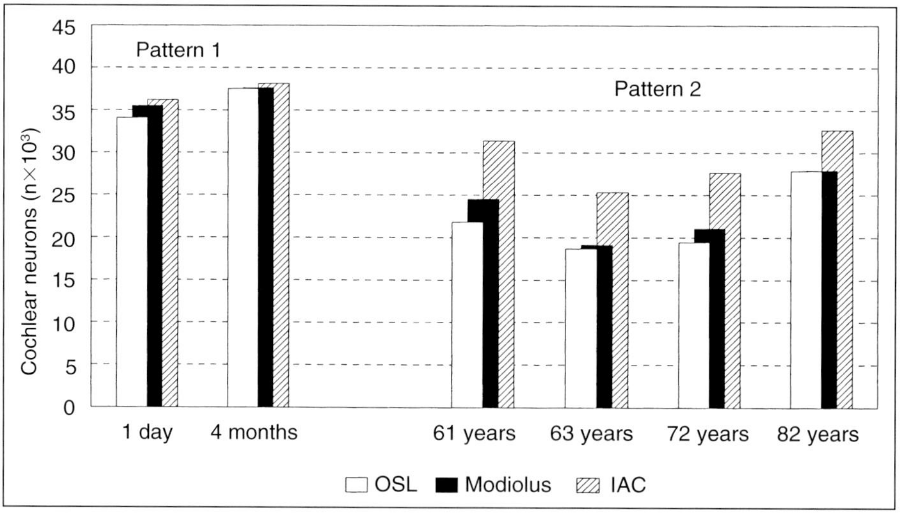

Fig. 4. Diagram showing total counts of cochlear neurons in the IAC, modiolus and OSL in the same cochlea of patients with ARNH. Degeneration pattern 1 was found in 2 very young children, whereas in patients over 60 years with ARNH degeneration pattern 2 was found.

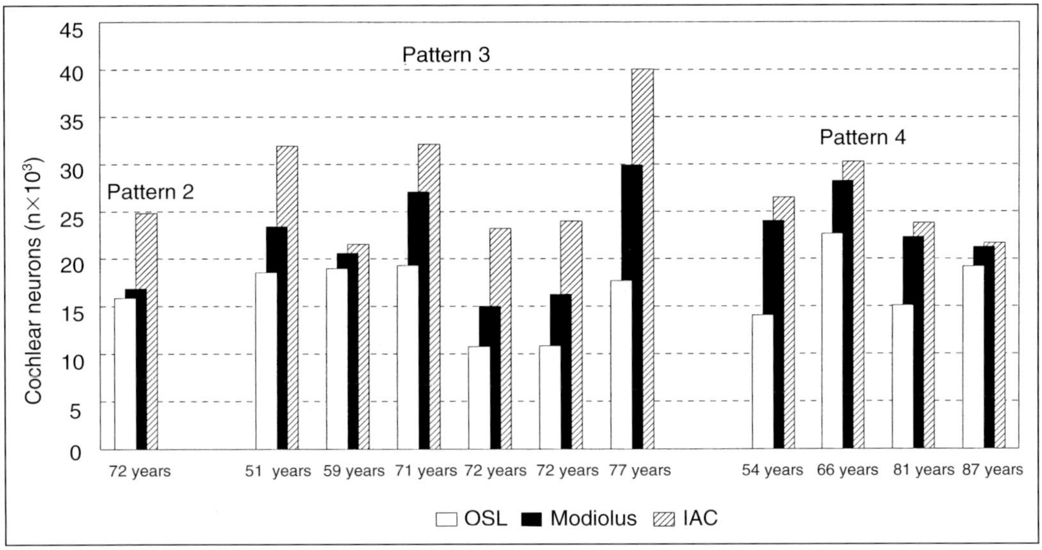

Fig. 5. Diagram showing total counts of cochlear neurons in the IAC, modiolus and OSL in the same cochlea of patients with SNHL. Only in 1 cochlea was degeneration pattern 2 present. The other cochleas revealed either degeneration pattern 3 or 4.

was also present in 4 cochleas of patients over 60 years with ARNH. In 6 cochleas, the counts of the SGCs were between the counts of the nerve fibers in the OSL and IAC and were defined as pattern 3. Pattern 4 showed similar counts of SGCs and nerve fibers in the IAC, whereas the counts in the OSL were much lower, which was found in 4 cochleas (fig. 5). In the group of patients with SNHL, 3 out of the 4 described degeneration patterns were found. No correlation between the severity of SNHL and degeneration pattern could be found.

Discussion

The quantitative analysis of cochlear neurons at 3 different sites in temporal bones from patients with either ARNH or SNHL confirmed and extended previously published data [15]. In all 45 cochleas from 25 patients, the loss of peripheral nerve fibers in the OSL was consistently greater than the loss of central nerve processes in the IAC. The difference in numbers varied between 2 and 74%.

The severer the nerve degeneration in the OSL, the larger the difference was between the numbers of central and peripheral fibers. Several investigators

[6–8, 16–20] have found that the loss of peripheral nerve processes in the OSL is consistently greater than the loss of SGCs or central nerve processes. The striking difference in the degeneration behavior of the peripheral and central nerve processes of the cochlear neurons would not have become apparent without a comparative quantitative study at the 3 sites of the same temporal bone.

In addition the counts of SGCs together with the counts in the OSL and IAC revealed 4 different degeneration patterns. Pattern 1 was only found in 2 cochleas from very young children showing similar high counts at all 3 sites which are highly suggestive of absent degeneration. In addition to pattern 1, pattern 2 was also observed in the group of patients with ARNH. The cochleas from the patients with SNHL showed in the majority degeneration pattern 3 or 4.

The present quantitative study of cochlear neurons in humans has shown that in young individuals the numbers of cochlear neurons are almost identical at all 3 sites. In patients over 60 years with ARNH, the loss of peripheral nerve processes is always severer than the loss of central nerve processes. This finding suggests that the central processes degenerate at a much slower rate or not at all.

The description of 4 different peripheral degeneration patterns shown in this study is somewhat preliminary due to the small number of cochleas. However, the results indicate that several degeneration patterns could occur. The factors responsible for the different degeneration patterns are still not understood and need further investigation.

References

1 Ylikoski J, Wersall J, Bjorkroth B: Degeneration of neural elements in the cochlea of the guinea-pig after damage to the organ of Corti by ototoxic antibiotics. Acta Otolaryngol Suppl (Stockh) 1974;326:23–41.

2 Kiang NY, Liberman MC, Levine RA: Auditory-nerve activity in cats exposed to ototoxic drugs and high-intensity sounds. Ann Otol Rhinol Laryngol 1976;85:752–768.

3 Leake PA, Hradek GT: Cochlear pathology of long term neomycin induced deafness in cats. Hear Res 1988;33:11–33.

4 Bichler E, Spoendlin H, Rauchegger H: Degeneration of cochlear neurons after amikacin intoxication in the rat. Arch Otorhinolaryngol 1983;237:201–208.

5 Spoendlin H: Factors inducing retrograde degeneration of the cochlear nerve. Ann Otol Rhinol Laryngol Suppl 1984;112:76–82.

6 Suzuka Y, Schuknecht HF: Retrograde cochlear neuronal degeneration in human subjects. Acta Otolaryngol Suppl (Stockh) 1988;450:1–20.

7 Felix H, Johnsson LG, Gleeson M, Pollak A: Quantitative analysis of cochlear sensory cells and neuronal elements in man. Acta Otolaryngol Suppl (Stockh) 1990;470:71–79.

8 Nadol JB Jr: Degeneration of cochlear neurons as seen in the spiral ganglion of man. Hear Res 1990;49:141–154.

9 Spoendlin H, Balle V, Bock G, Bredberg G, Danckwardt LN, Felix H, Gleeson M, Johnsson LG, Luciano L, Rask AH, et al: Multicentre evaluation of the temporal bones obtained from a patient with suspected Ménière's disease. Acta Otolaryngol Suppl (Stockh) 1992;499:1–21.

10 Felder E, Kanonier G, Scholtz A, Rask AH, Schrott FA: Quantitative evaluation of cochlear neurons and computer-aided three-dimensional reconstruction of spiral ganglion cells in humans with a peripheral loss of nerve fibres. Hear Res 1997;105:183–190.
11 Spoendlin H, Schrott A: The spiral ganglion and the innervation of the human organ of Corti. Acta Otolaryngol (Stockh) 1988;105:403–410.
12 Spoendlin H, Schrott A: Analysis of the human auditory nerve. Hear Res 1989;43:25–38.
13 Spoendlin H, Schrott A: Quantitative evaluation of the human cochlear nerve. Acta Otolaryngol Suppl (Stockh) 1990;470:61–69.
14 Hawkins JE Jr, Johnsson LG, Stebbins WC, Moody DB, Coombs SL: Hearing loss and cochlear pathology in monkeys after noise exposure. Acta Otolaryngol (Stockh) 1976;81:337–343.
15 Felix H, Gleeson MJ, Pollak A, Johnsson LG: The cochlear neurons in humans; in Iurato S, Veldman JE (eds): Progress in Human Auditory and Vestibular Histopathology. Amsterdam, Kugler, 1997, pp 73–79.
16 Johnsson LG, Hawkins JE: Sensory and neural degeneration with aging, as seen in microdissections of the human inner ear. Ann Otol Rhinol Laryngol 1972;81:179–193.
17 Johnsson LG: Cochlear anatomy and histopathology; in Grey R (ed): Cochlear Implants. London, Croom Helm, 1985, pp 50–73.
18 Pollak A, Felix H: Histopathological features of the spiral ganglion and cochlear nerve in temporal bones from three patients with profound hearing loss. Acta Otolaryngol Suppl (Stockh) 1985; 423:59–66.
19 Nadol JB Jr, Young YS, Glynn RJ: Survival of spiral ganglion cells in profound sensorineural hearing loss: Implications for cochlear implantation. Ann Otol Rhinol Laryngol 1989;98:411–416.
20 Nadol JB Jr, Xu WZ: Diameter of the cochlear nerve in deaf humans: Implications for cochlear implantation. Ann Otol Rhinol Laryngol 1992;101:988–993.

Dr. Heidi Felix, ENT Department, University Hospital of Zürich,
Frauenklinikstrasse 24, CH–8091 Zürich (Switzerland)
Tel. +41 1 225 58 05, Fax +41 1 225 89 59, E-Mail felix@orl.usz.ch

Felix D, Oestreicher E (eds): Rational Pharmacotherapy of the Inner Ear.
Adv Otorhinolaryngol. Basel, Karger, 2002, vol 59, pp 124–130

The Inner Hair Cell Afferent/Efferent Synapses Revisited: A Basis for New Therapeutic Strategies

Jean-Luc Puel, Jérôme Ruel, Matthieu Guitton, Rémy Pujol

Laboratoire de Neurobiologie de l'Audition – Plasticité synaptique,
INSERM UMR 254, Université Montpellier-1, Montpellier, France

Abstract

Within the cochlea, the sensory inner hair cells, which transduce the mechanical displacement of the basilar membrane into neural activity, release glutamate that acts on post-synaptic α-amino-3-hydroxy-5-methyl-4-isoxazolepropionate (AMPA) receptor channels located on dendrites of primary auditory neurons. Up to now, it has been thought that the auditory nerve responses passively reflected the motion of the basilar membrane supporting the organ of Corti. Here, we show that dopaminergic lateral olivocochlear efferents drive a permanent gain control at the site of auditory action potential initialization. A dysfunction of this system leads to the development of early signs of excitotoxicity. With the knowledge of the molecular mechanisms involved at this first synaptic complex in the cochlea, it is now possible to envisage local treatments for spiral ganglion neurons, either to stop an excitotoxically induced hyperexcitability (probably the starting point of most posttraumatic tinnitus) or to prevent neuronal death (neural presbycusis).

The inner hair cells (IHCs) are the mechanoelectrical transducers of the inner ear. They synapse with radial dendrites of the spiral ganglion neurons and the auditory message is conveyed to the cochlear nucleus through the auditory nerve (fig. 1). On the other side, the outer hair cells (OHCs), through a reverse electromechanical transduction (active mechanism), enhance the motion of the basilar membrane and give IHC transduction its exquisite sensibility. Two types of efferent systems provide a feedback modulation from the central nervous system to these two types of sensory cells [1]. A lateral efferent system originating

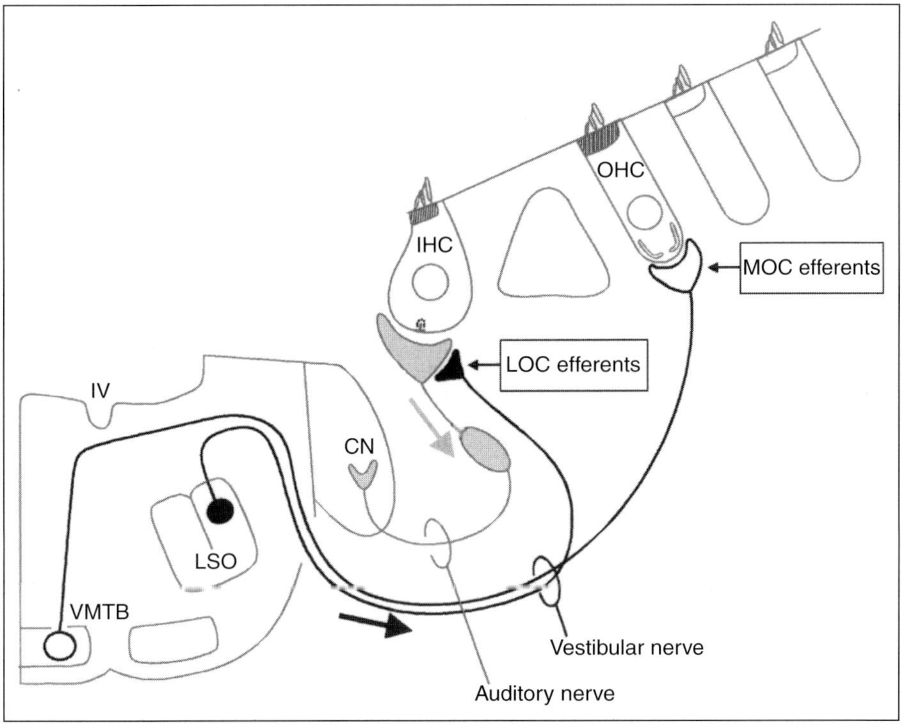

Fig. 1. Schematic representation of afferent/efferent feedback loops in the organ of Corti. The IHCs are synaptically connected to the radial dendrites of the spiral ganglion neurons. These neurons are connected to the cochlear nucleus (CN) through the auditory nerve neurons. In turn, the brainstem lateral superior olivary nucleus (LSO) neurons project to the auditory dendrites below IHCs, forming what is called lateral olivo-cochlear (LOC) efferents. The outer hair cells (OHCs) are directly innervated by endings of neurons from the ventromedial nuclei of the trapezoid body (VMTB) forming the medial olivocochlear (MOC) efferents. Both efferent systems run within the vestibular nerve up to the entrance of the cochlea. The spiral afferent system from OHCs to the cochlear nucleus is not represented. Bold arrows indicate the direction of action potential propagation along the neurons. IV = Floor of 4th ventricle.

from the lateral superior olive projects on auditory nerve dendrites underneath IHCs. A medial olivocochlear system, originating from medial nuclei of the superior olivary complex, directly connects OHCs.

Glutamate (Glu) is the main accepted neurotransmitter at the IHC-auditory nerve synapse [2]. Based on chemical neuroanatomy results [3], the current knowledge on efferent neurotransmitters can be summarized as follows. The lateral efferents may use several neurotransmitters or neuromodulators [acetylcholine, γ-aminobutyric acid, dopamine (DA), enkephalins, dynorphins and calcitonin gene-related peptide]. Similarly, a coexistence of acetylcholine, γ-aminobutyric acid and/or calcitonin gene-related peptide in medial efferent neurons has been proposed. To date, most of the functional published results only concern the effects of acetylcholine on the OHC electromotility.

This report summarizes the most recent neuropharmacological data at the IHC afferent/efferent synaptic complex: the type of Glu receptors involved and their modulation by the lateral efferent neurotransmitters. These new findings allow us to envisage new therapeutic local strategies, either to protect spiral ganglion neurons against excitotoxic injury (traumatic and/or ischemic sudden deafness), to prevent an excitotoxically induced hyperexcitability (probably the starting point of most posttraumatic tinnitus) or to delay neuronal death (neural presbycusis).

Pharmacology of the Glutamatergic Synapse

The ionotropic Glu receptors, generally used for fast excitatory synaptic transmission, are classically divided into three types of receptors named after their sensitivity to agonists: N-methyl-*D*-aspartate (NMDA), α-amino-3-hydroxy-5-methyl-4-isoxazolepropionate (AMPA) and kainate. Analysis of ionotropic Glu receptors with gene expression, immunocytochemistry and in situ hybridization indicates that primary auditory nerve cells express NMDA (NR1 and NR2A–D), AMPA (GluR2–4), kainate (GluR5–7) receptor subunits as well as the high-affinity kainate-binding proteins (KA1 and KA2) [4–8]. This suggests that NMDA, AMPA and kainate receptors might coexist on primary auditory nerve cells. However, until recently, the role of each of these receptors at the IHC-auditory nerve synapse has been controversial and unsettled.

Up to now, the lack of a specific antagonist that differentiates between AMPA and kainate receptor-evoked responses has limited our ability to determine the specific involvement of each of these two receptors. Recently, GYKI 53784 (LY303070) has been demonstrated to be one of the most selective antagonists for AMPA receptors [9]. Taking advantage of this new pharmacological tool, the role of AMPA receptors in the fast synaptic transmission and excitotoxicity was addressed. We compared GYKI 53784 (LY303070) with additional AMPA/kainate antagonists, GYKI 52466 and 6,7-dinitroquinoxaline-2,3-dione, and the NMDA antagonist, *D*-2-amino-5-phosphonopentanoate, in several electrophysiological (fig. 2), neurotoxicological and histochemical tests [10, 11].

Whereas the NMDA antagonist *D*-2-amino-5-phosphonopentanoate had no effect, GYKI 53784 had the same potency as 6,7-dinitroquinoxaline-2,3-dione

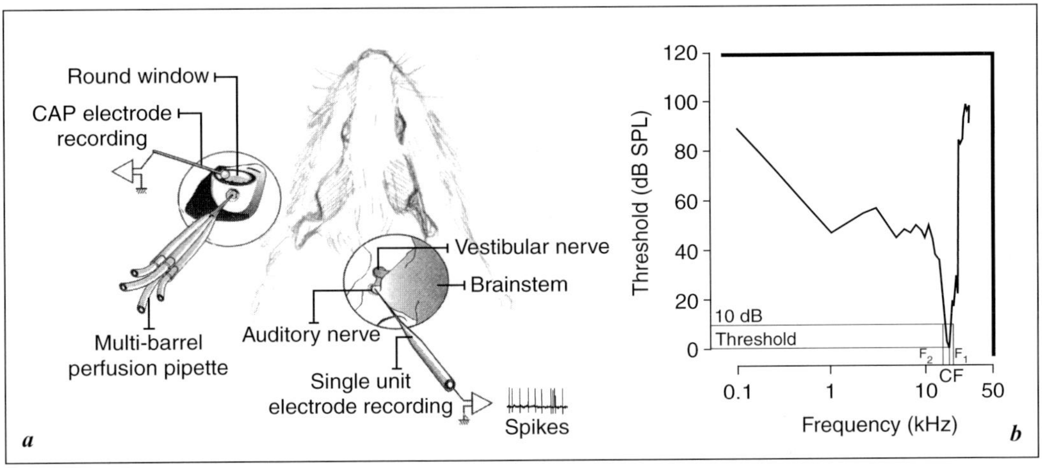

Fig. 2. Schematic representation of the design of the experiment. ***a*** The perilymphatic perfusion technique. The cochlea was exposed using a dorsal approach (left side of the figure). A 0.5-mm hole was gently drilled into the scala tympani of the basal turn of the cochlea to receive a multi-barrel perfusion pipette (ASI Instruments) and a compound action potential (CAP) recording electrode placed on the round window of the cochlea. The test solutions were allowed to flow out of the cochlea through a hole (0.2 mm diameter) made at the apex. The cochlear nerve was exposed using a posterior fossa approach (right side of the figure). The unit activity was tracked by advancing the microelectrode through the cochlear nerve with a motorized micromanipulator (Micro-control, module 80) during exposure to 80 dB SPL white noise generated by a Bruël & Kjaer (type 1405; bandwidth 100 kHz). Once a single unit had been isolated, spontaneous activity was averaged during 10 s. ***b*** A computer-controlled threshold-tracking program using a 200-ms tone burst presented 3/s then obtained the single unit tuning curves. The threshold criterion was a difference of 10 spikes/s, i.e. 2 spikes between the tone (200 ms) and nontone (200 ms) counting intervals. The program determined the characteristic frequency (CF) and the frequency (F) tuning of the fiber, by measuring the $Q_{10\,dB}$ defined as the characteristic frequency divided by the bandwidth at 10 dB above the characteristic frequency threshold.

in reducing the compound action potential. To determine the mechanisms underlying the reduction of the compound action potential, the effect of GYKI 53784 on the spontaneous activity of the single auditory nerve fibers was studied. Perfusion of 10 μ*M* GYKI 53784 drastically reduced the spontaneous discharge rate of the auditory nerve fibers. In cases where the fiber was held for a sufficient amount of time, the drug was washed out and the same paradigm was repeated with 50 and 100 μ*M* of GYKI 53784. The activity of the fiber was completely abolished by 50 or 100 μ*M* of GYKI 53784, suggesting that fast excitatory neurotransmission between the IHCs and the primary auditory nerve fibers is predominantly mediated by AMPA-preferring receptors.

Conversely, the NMDA antagonist *D*-2-amino-5-phosphonopentanoate had no effect on cochlear potentials. This result is in agreement with the data of Nakagawa et al. [12], showing the absence of NMDA responses in guinea pig spiral ganglion neuron somata isolated in the same manner as the chicken auditory nerve somata which demonstrated an NMDA response. Consistently with these negative data, several authors have reported that, when applied through the perilymphatic space of the cochlea in vivo, NMDA and NMDA antagonists induce a very small or no response [13].

Modulation of the Glutamatergic Synapse by the Lateral Efferents

While the physiology of the medial olivocochlear efferents has been extensively studied, little is known about the functional significance of lateral olivocochlear efferents. The role of the latter is still unclear because it is difficult to selectively stimulate or destroy this system. Due to these technical limitations, it was decided to use an in vivo pharmacological approach utilizing DA, one of the neurotransmitters used by lateral olivocochlear efferents [2, 3].

When perfused into the cochlea, 1 m*M* DA induced a reduction in spontaneous firing. This effect was completely reversed by washing DA out of the cochlea with control artificial perilymph. Although DA did not affect the fibers' frequency tuning responses to tones, it did elevate the threshold at and off the characteristic frequency. This suggests that the unit threshold elevations induced by DA perfusion are not linked to changes in tuning properties of the auditory fibers. When stimulated at the characteristic frequency, DA decreased the sound-driven activity and reduced the dynamic discharge range of the fibers.

To assess the role of endogenous DA, we used specific DA antagonists. As described previously by Liberman [14], at least two classes of fibers were found, those with low spontaneous rates ($<$20 spikes/s) and those with high spontaneous rates ($>$20 spikes/s). In the low spontaneous rate fibers, DA blockade resulted in an increase in the spontaneous firing rate with little or no postexcitatory suppression. This increase in basal activity of the fibers resulted in an improvement of the sensitivity at threshold and a reduction in the dynamic range. In high spontaneous rate fibers, DA blockade induced a brief increase in firing rate, immediately followed by a reduction to values below predrug rates.

To investigate the mechanism underlying the postexcitatory inhibition, cochleas perfused with eticlopride were fixed and processed for transmission electron microscopy. While no abnormality could be detected in the cochlea perfused with artificial perilymph, a clear swelling of some of the afferent dendrites connected to the inner hair cells was observed in the cochlea perfused with the DA antagonist eticlopride. The swollen radial dendrite terminals exhibited disarray of the cytoskeleton. IHCs contained normal presynaptic elements in synapses with swollen radial dendrites. Similarly, normal-looking vesiculated efferents made synaptic contact with the swollen radial dendrites. This suggests that the marked reduction in firing rate observed on the higher preperfusion rate fibers may reflect early signs of excitotoxicity that occurred during DA blockade. Consistently with this assumption, GYKI 53784 blocked the eticlopride-induced swelling of radial dendrites. The mechanism by which the DA antagonist eticlopride facilitates glutamate excitotoxicity may be due to a relief of an inhibitory modulation of AMPA receptors by D_2 receptors located on the radial dendrites.

Conclusions

Up to now, it has been thought that the auditory nerve responses passively reflected the motion of the basilar membrane supporting the organ of Corti. Here we confirm that AMPA and not kainate or NMDA receptors mediate fast excitatory synaptic transmission. Moreover, the activity of AMPA receptors is tonically modulated by the dopaminergic lateral efferent system, which acts as a permanent gain control at the initialization site of the auditory action potential. This gain control is responsible for the maintenance of fundamental characteristics of auditory nerve responses. Dysfunction of this system leads to the development of early signs of Glu-induced excitotoxicity. In the near future we could well expect some treatment to help the regrowth of nerve fibers in humans, the protection and/or the repair of synapses and the functional recovery after ischemic and/or noise-induced hearing loss (sudden deafness). Similarly, it may be possible to envisage treatments for spiral ganglion neurons, either to stop an excitotoxically induced hyperexcitability (probably the starting point of most posttraumatic tinnitus) or to prevent neuronal death (neural presbycusis). Based upon the experimental results reported here and in related papers, a new therapeutic strategy should be developed: a local application of drugs, as via a transtympanic catheter. This new pharmacological strategy allows to precisely target the molecular mechanism involved and to use concentrations low enough to avoid side effects.

References

1 Warr WB: Organisation of olivocochlear efferent systems in mammals; in Webster DB, Popper AN, Fay RR (eds): The Anatomy of Mammalian Auditory Pathway. Berlin, Springer, 1992, pp 410–448.
2 Puel JL: Chemical synaptic transmission in the cochlea. Prog Neurobiol 1995;447:449–476.

3 Eybalin M: Neurotransmitters and neuromodulators of the mammalian cochlea. Physiol Rev 1993; 73:309–373.
4 Ryan AF, Brumm D, Kraft M: Occurrence and distribution of non-NMDA glutamate receptor mRNAs in the cochlea. Neuroreport 1991;2:643–646.
5 Safieddine S, Eybalin M: Triple fluorescence evidence for the coexistence of acetylcholine, enkephalins and calcitonin gene-related peptide within efferent (olivocochlear) neurons of rats and guinea-pigs. Eur J Neurosci 1992;4:981–992.
6 Niedzielski AS, Wenthold RJ: Expression of AMPA, kainate, and NMDA receptor subunits in cochlear and vestibular ganglia. J Neurosci 1995;15:2338–2353.
7 Usami S, Matsubara A, Fujita S, Shinkawa H, Hayashi M: NMDA (NMDAR1) and AMPA-type (GluR2/3) receptor subunits are expressed in the inner ear. Neuroreport 1995;6:1161–1164.
8 Matsubara A, Laake JH, Davanger D, Usami S, Ottersen OP: Organization of AMPA receptor subunits at a glutamate synapse: A quantitative immunogold analysis of hair cell synapses in the rat organ of Corti. J Neurosci 1996;16:4457–4467.
9 Bleakman D, Bally BA, Schoepp DD, Palmer AJ, Bath CP, Sharpe EF, Woolley ML, Bufton HR, Kamboj RK, Tamawa I, Lodge D: Activity of 2,3-benzodiazepines at native rat and recombinant human glutamate receptors in vitro: Stereospecificity and selectivity profiles. Neuropharmacology 1996;35:1689–1702.
10 Ruel J, Chen C, Pujol R, Bobbin RP, Puel JL: AMPA-preferring glutamate receptors in cochlear physiology of adult guinea-pig. J Physiol (Lond) 1999;518:667–680.
11 Ruel J, Bobbin RP, Vidal D, Pujol R, Puel JL: The selective AMPA receptor antagonist GYKI 53784 blocks action potential generation and excitotoxicity in the guinea pig cochlea. Neuropharmacology 2000;39:1959–1973.
12 Nakagawa T, Komune S, Uemura T, Akaike N: Excitatory amino acid response in isolated spiral ganglion cells of guinea pig cochlea. J Neurophysiol 1991;65:715–723.
13 Bobbin RP, Bledsoe SC Jr, Jenison GL: Neurotransmitters of the cochlea and the lateral line organ; in Berlin CI (ed): Hearing Sciences: Recent Advances. San Diego, College Hill Press, 1984, pp 159–180.
14 Liberman MC: Auditory-nerve response from cats raised in a low-noise chamber. J Acoust Soc Am 1978;63:442–455.

Dr. Jean-Luc Puel, INSERM U-254, Laboratoire de Neurobiologie de l'Audition,
71, rue de Navacelles, F–34090 Montpellier (France)
Tel. +33 467 417 716, Fax +33 467 417 716, E-Mail puel@montp.inserm.fr

Felix D, Oestreicher E (eds): Rational Pharmacotherapy of the Inner Ear.
Adv Otorhinolaryngol. Basel, Karger, 2002, vol 59, pp 131–139

Neurotransmission of the Cochlear Inner Hair Cell Synapse – Implications for Inner Ear Therapy

Elmar Oestreicher[a, b], *Wolfgang Arnold*[a], *Dominik Felix*[b]

[a] ENT Department, Technical University of Munich, Germany;
[b] Division of Neurobiology, University of Berne, Switzerland

Abstract

The cochlear inner hair cells (IHCs) are connected to afferent type I auditory neurons and use probably *L*-glutamate as a neurotransmitter. This IHC synapse receives efferent input from the lateral part of the efferent olivocochlear system with neurons originating in the brainstem and terminating below IHCs synapsing with the afferent type I dendrites. A number of substances have been proposed to function as neurotransmitter or neuromodulator in the lateral efferent system: acetylcholine, γ-aminobutyric acid (GABA), dopamine, enkephalin and dynorphin. With the aid of microiontophoretic techniques, we studied several transmitter candidates and characterized their receptor subtypes as well as their function on spontaneous or evoked activity of afferent dendrites. The results showed that the glutamatergic transmission of IHCs is facilitated by all types of glutamate receptors: ionotropic glutamate receptors of the N-methyl-*D*-aspartic acid (NMDA) and α-amino-3-hydroxy-5-methyl-4-isoxazolepropionic acid (AMPA) type as well as group I and II metabotropic glutamate receptors. This excitatory glutamatergic transmission is under inhibitory control of GABA (mediated by $GABA_A$ receptors) and dopamine (mediated by D_1 and D_2 receptors). In contrast, acetylcholine was able to excite afferent dendrites via muscarinic receptors. These results demonstrate that the lateral efferent system has modulatory function on the glutamatergic neurotransmission of IHCs. Excitation of afferent dendrites by glutamate released from IHCs can thus be tuned in different physiological or pathophysiological conditions. This could have therapeutic implications as it is known that noise exposure is followed by an excitotoxic injury of the IHC synapse. During overexcitation of IHCs, a possible therapy based on the neurochemical data would be (a) glutamate antagonists, (b) dopamine agonists, (c) GABA agonists or a combination from a, b and a, c.

The nerve fiber system within the mammalian cochlea is divided into two main classes: an afferent and an efferent system. The afferent innervation connecting the cochlear hair cells to the cochlear nucleus complex is responsible for coding the auditory messages. 95% of the spiral ganglion neurons, the type I auditory neurons, connect the inner hair cells (IHCs) to the cochlear nuclei, responsible for coding the auditory messages. 5% of the spiral ganglion neurons, the type II auditory neurons, connect the outer hair cells (OHCs) to the cochlear nucleus complex. The function of these neurons is not known.

The brainstem sends its nerve fibers to the auditory receptor via the efferent olivocochlear system, involved in the modulation of auditory messages. The efferent innervation can also be divided into two parts: a medial and a lateral part. The medial efferent system is projecting directly on OHCs, modulating the motility of the OHCs. The lateral efferent system projects with axodendritic synapses on IHC afferents. This system seems to modulate the auditory message.

The afferent neurotransmitter between IHCs and afferent nerve endings in the cochlea seems to be *L*-glutamate [for a review, see 1].

For the lateral efferent innervation of the cochlea, a number of substances have been proposed to have neurotransmitter or neuromodulator function: acetylcholine, γ-aminobutyric acid (GABA), dopamine, enkephalin, dynorphin and calcitonin gene-related peptide [1]. This enumeration illustrates the complex regulation of the IHC synapse in the cochlea by different neurotransmitters and their receptors. This highly organized synapse at the IHC level seems to be necessary to facilitate normal hearing function. Moreover, as neurotransmitters and their receptors are also involved in pathological situations like noise trauma, ischemia [2] or tinnnitus, this knowledge may provide therapeutic strategies to influence the course of inner ear diseases.

In this study, we used microiontophoretic techniques to investigate the function of several neurotransmitter candidates in vivo in the guinea pig. We chose the major neurotransmitter candidates glutamate, GABA, dopamine and acetylcholine. In addition we investigated the function of the neurotrophic factor NT-3 on the neurotransmission of IHCs. NT-3 is expressed in the mammalian cochlea and it is known that neurotrophic factors maintain normal synaptic transmission.

Methods

The experiments were carried out on pigmented female guinea pigs (Novartis; weight 330–450 g). For premedication, atropine (0.5 mg/kg) was injected intramuscularly 30 min prior to anesthesia. Animals were anesthetized with a combination of ketamine HCl (Ketalar 100; 1.2 ml/kg, Gräub, Berne, Switzerland) and Rompun (0.4 ml/kg; Bayer, Leverkusen, Germany). To ensure a constant level of anesthesia, additional low doses were administered regularly. After tracheotomy for artificial respiration, the auditory bulla was approached laterally and the left cochlea exposed. A small opening was drilled into the cochlear bone over the pigmented stria vascularis and the spiral ligament of the third turn. Multibarreled microelectrodes with a tip diameter of 2 μm were inserted through the basal border of the stria vascularis and then driven almost parallel to the tectorial membrane. The subsynaptic region was reached at a depth of about 200–280 μm as indicated by typical phasic activity. Extracellular action potentials were recorded by using a 2 *M* NaCl-filled barrel of the microelectrode. Substances applied microiontophoretically with appropriate anionic or cationic currents by the other four channels of the microelectrode included: NMDA (N-methyl-*D*-aspartic acid; Sigma; 0.1 *M*, pH 7.5, adjusted with NaOH), AMPA (α-amino-3-hydroxy-5-methylisoxazole-4-propionic acid; Tocris; 0.01 *M*, pH 7.5, NaOH), CNQX (6-cyano-7-nitroquinoxaline-2, 3-dione; RBI; 1 m*M*, pH 8.5), *L*-glutamic acid (Sigma; 0.1 *M*, pH 8.5), AIDA (RS-1-amino-indan-1,5-dicarboxylic acid; Tocris; 0.0044 *M*, pH 10.5, NaOH), DHPG (S-3,5-dihydroxyphenylglycine; Tocris; 0.005 *M*, pH 8.5, NaOH), dopamine (Sigma; 0.2 *M*, pH 4, adjusted with HCl), Cl-APB (6-chloro-7,8-dihydroxy-3-allyl-1-phenyl-2,3,4,5-tetrahydro-1H-3-benzazepine hydrobromide; RBI; 0.02 *M*, pH 4,3, adjusted with HCl), bromocriptine (bromocriptine methanesulfonate; RBI; 0.01 *M*, pH 3.5, adjusted with HCl), GABA (Fluka; 0.5 *M*, pH 3–3.5, adjusted with HCl), bicuculline methiodide (Sigma; 5 m*M*, pH 3, adjusted with HCl), muscimol (5-aminomethyl-3-hydroxyisoxazole; Sigma; 5 m*M*, pH 3–3.5, adjusted with HCl), baclofen (4-amino-3-[4-chlorophenyl]butanoic acid; Sigma; 10 m*M*, pH 3–3.5, adjusted with HCl), acetylcholine chloride (0.5 *M*, pH 3.5, adjusted with HCl), NT-3 (Regeneron; 20 n*M* in 0.1% BSA, PBS), AP-7 (*D*-2-aminophosphonoheptanoate; Novartis; 0.2 *M*, pH 8, adjusted with NaOH), atropine (Sigma; 0.7 *M*, pH 3.8, adjusted with HCl), *D*-tubocurarine (Sigma; 1.1 *M*, pH 4.5, adjusted with HCl).

Results

In a first set of experiments, we determined the type of glutamate receptors involved in the neurotransmission of IHCs.

Iontophoretically applied *L*-glutamate as well as the ionotropic glutamate receptor (iGluR) agonists NMDA and AMPA increased the firing reversibly in all tested afferent fibers of IHCs. Application of the selective metabotropic glutamate receptor (mGluR) agonist DHPG, ejected with similar currents and equal periods of time, increased the activity of afferent dendrites of IHCs as well. However, compared to glutamate, NMDA or AMPA, the DHPG-induced activation showed a significantly longer latency of onset and recovery as well as a longer duration of the stimulating effect (fig. 1a). The DHPG-induced activity could completely be antagonized by the specific mGluR antagonist AIDA in a reversible manner, while the AMPA-induced activity could only be blocked by the specific AMPA antagonist CNQX but not by the specific mGluR antagonist AIDA (fig. 1b).

Secondly, the influences of the neurotransmitters GABA, dopamine and acetylcholine on the glutamatergic neurotransmission of IHCs were tested.

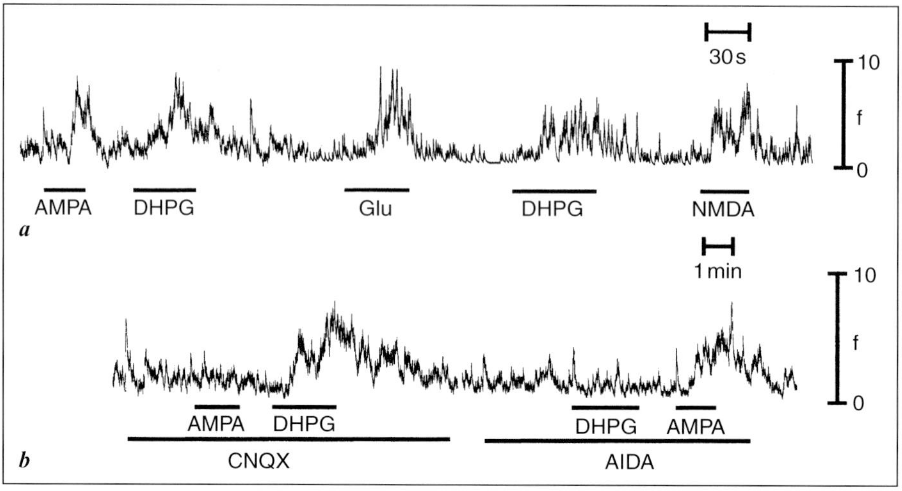

Fig. 1. Effect of ionotropic and metabotropic glutamate receptor agonists and antagonists on the activity of a spontaneous firing fiber: *L*-glutamate and the ionotropic agonists AMPA and NMDA as well as the metabotropic agonist DHPG increase the firing of an afferent neuron (***a***). The effect of AMPA can be blocked by the AMPA receptor antagonist CNQX, while the effect of DHPG is not influenced by CNQX (***b***). The metabotropic glutamate receptor antagonist AIDA is able to block the effect of DHPG but not that of AMPA. Fiber activity is demonstrated on integrated time frequency curves. Administration of the substances is indicated by bars beneath the frequency curves; f = frequency in hertz (Hz).

The activated firing (by application of AMPA and NMDA) was inhibited by the simultaneous application of GABA (fig. 2a). Muscimol, a specific $GABA_A$ receptor agonist, and baclofen, a specific $GABA_B$ receptor agonist, were applied to the activated fibers. Muscimol reflected the inhibitory effect of GABA whereas baclofen did not attenuate the activated discharge in any neuron (fig. 2a). Moreover, bicuculline as an antagonist of the $GABA_A$ subtype blocked the effect of GABA which was shown by the recovery of the activated fiber discharge.

When tested on activated fibers (NMDA or AMPA), dopamine showed also an inhibitory action: dopamine reduced the induced firing rate of afferent neurons of IHCs in a dose-dependent manner (fig. 2b).

The iontophoretic application of Cl-APB as an agonist of the D_1-like receptor subtype and bromocriptine as a D_2-like receptor subtype preferring agonist mimicked the effects of dopamine and showed in all units tested an inhibitory effect on the activated fibers (fig. 2b). The degree of effectiveness and duration of action of both agonists on activated nerve fibers were comparable and showed no significant differences between D_1 and D_2 agonists. Spontaneous nerve activity could not be influenced by either D_1 or D_2 agonists.

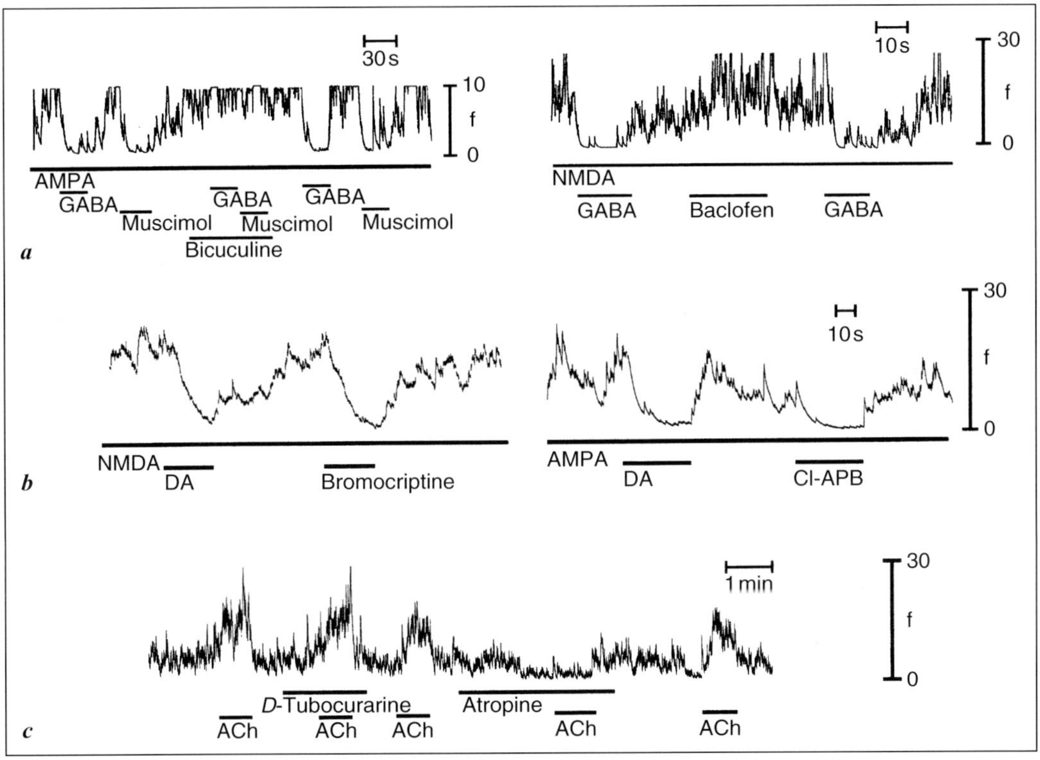

Fig. 2. Effect of GABA, dopamine and acetylcholine on the activity of afferent neurons. Fiber activity is demonstrated on integrated time frequency curves. Administration of the substances is indicated by bars beneath the frequency curves; f = frequency in hertz (Hz). ***a*** GABA as well as muscimol, a specific $GABA_A$ receptor agonist, show inhibitory effects on an activated (AMPA) afferent fiber. Application of a $GABA_A$ receptor antagonist (bicuculline) suppresses inhibition of the activated afferent fiber by both GABA and muscimol. There is no inhibition of the activated (NMDA) afferent fiber by the specific $GABA_B$ receptor agonist (baclofen). ***b*** Recording of two different fibers showing an activated fiber (NMDA) during application of dopamine (DA) and bromocriptine and an activated fiber (AMPA) during application of dopamine and Cl-APB. All substances showed inhibitory effects. ***c*** Acetylcholine (ACh) and the muscarinic agonist atropine could increase the firing rate of an afferent neuron. The application of the nicotinic drug *D*-tubocurarine was without any effect.

The action of acetylcholine was different compared to that of dopamine or GABA. The iontophoretic application of acetylcholine increased the firing of afferent neurons. This effect is probably due to muscarinic receptor subtypes: *D*-tubocurarine, a nicotinic antagonist, could not block the effect of acetylcholine while atropine, a muscarinic antagonist, inhibited the effects of acetylcholine (fig. 2c).

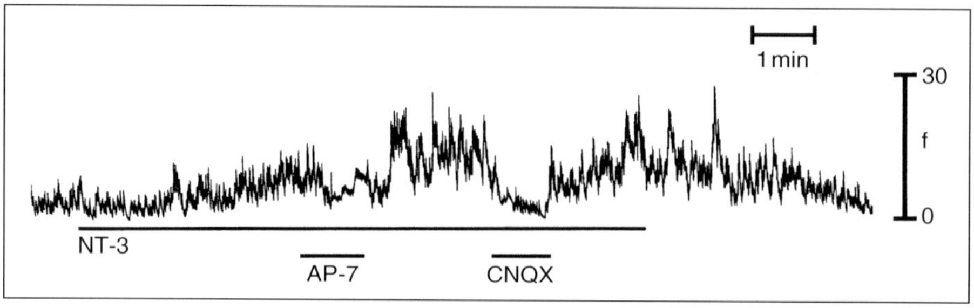

Fig. 3. Effect of NT-3 and its blockade by AP-7 and CNQX on a spontaneously firing afferent fiber of an IHC. Both glutamate antagonists AP-7 and CNQX are able to block the NT-3-induced activity. Fiber activity is demonstrated on integrated time frequency curves. Administration of the substances is indicated by bars beneath the frequency curves; f = frequency in hertz (Hz).

Finally we tested the action of the neurotrophin NT-3 on spontaneous firing rates of afferent synapses of IHCs (fig. 3). Application of NT-3 in a dose of 100 nA for periods of about 10 min led to an increase in fiber activity with a gradual return to the baseline activity after cessation of the application of NT-3. In addition, the neurotrophin-induced discharge could be blocked by the specific ionotropic NMDA and AMPA glutamate receptor antagonists AP-7 (NMDA antagonist) and CNQX (AMPA antagonist).

Discussion

Glutamate is the physiological mediator of most excitatory synaptic transmission in the central nervous system. The action of glutamate is mediated by either ionotropic (receptors coupled to a ion channel, iGluR) or metabotropic receptors (receptors coupled to a G protein, mGluR). During excessive stimulation of postsynaptic receptors, the physiological neurotransmitter glutamate exerts neurotoxic action in the nervous system (excitotoxicity) [3]. Different pathophysiological conditions such as ischemia, hypoglycemia, anoxia and trauma have been linked to excitotoxicity [3].

The cochlear IHCs use probably *L*-glutamate as the afferent neurotransmitter [1, 4]. As excitotoxicity contributes to different inner ear diseases [2, 5] like noise trauma, progressive hearing loss and tinnitus, characterization of the neurotransmitters and receptors at the IHC synapse is necessary to develop a neuropharmacological treatment of these disorders.

Regarding the glutamate receptors we could demonstrate the existence of iGluR of the NMDA and the AMPA type (fig. 1a). The contribution of the NMDA receptors to sound processing is still controversial [6], but immunohistochemical, gene expression and in situ hybridization studies confirm the presence of subunits that present the NMDA (NMDAR1) and the AMPA (GluR2/3) receptor [7].

Moreover, we could demonstrate the presence of mGluR (fig. 1a, b) [8]. The activation of afferent neurons by the group I mGluR agonist DHPG is in good accordance with recent immunohistochemical and in situ hybridization studies indicating the existence of group I mGluR on primary afferent neurons and cochlear hair cells [9, 10].

Microiontophoretic investigation of the neurotransmitters of the efferent axodendritic innervation of IHC afferents revealed an even more complex situation.

As reported earlier, dopamine showed an inhibitory effect on the afferent neurons, mediated via both D_1- and D_2-like receptor subtypes (fig. 2b) [11]. These findings could be confirmed recently on isolated spiral ganglion cells [12]. A similar inhibitory effect could be observed with GABA (fig. 2a). GABA seems to act through $GABA_A$ receptor subtypes (fig. 2a) [13]. In contrast, acetylcholine exhibited an excitatory effect that could be due to muscarinic receptors, as atropine could block this effect (fig. 2c). This is in accordance with in situ hybridization, and PCR studies showed M_3 muscarinic receptors postsynaptically to IHCs [14].

Surprisingly the microiontophoretic application of the neurotrophic factor NT-3 led to a long-lasting increase in the discharge rate of afferent neurons (fig. 3) [15]. We observed that both ionotropic glutamate receptors NMDA and AMPA are involved in the measured NT-3 effect as the increased activity could be blocked by AP-7 and CNQX. This may indicate a direct effect of NT-3 on the transmitter release of glutamate from the IHC presynapse or the NT-3 response is mediated by trkC receptors (the specific receptors of NT-3) postsynaptically to IHCs via modulation of postsynaptic glutamate receptor sensitivity. The more slowly and long-lasting effect of NT-3 could be responsible for maintaining a continuous transmission between IHCs and afferent neurons. Depending on the status of the synapses, NT-3 could alter the transmission via NMDA and AMPA receptors to a more or less excitable status.

Thus, NT-3 could act as an intrinsic amplifier of low noise sound levels or a trophic factor involved in sensitization of habituation to distinct sound levels.

Taken together we can summarize that the glutamatergic neurotransmission between IHCs and afferent dendrites is mediated via iGluR of the NMDA and AMPA type and via group I mGluR. This afferent transmission is controlled by the lateral part of the efferent olivocochlear system using GABA, dopamine and acetylcholine. GABA acting via $GABA_A$ receptor subtypes and dopamine acting via D_1- and D_2-like receptors exhibit an inhibitory influence on the activity of afferent neurons. Acetylcholine seems to facilitate an excitatory effect via

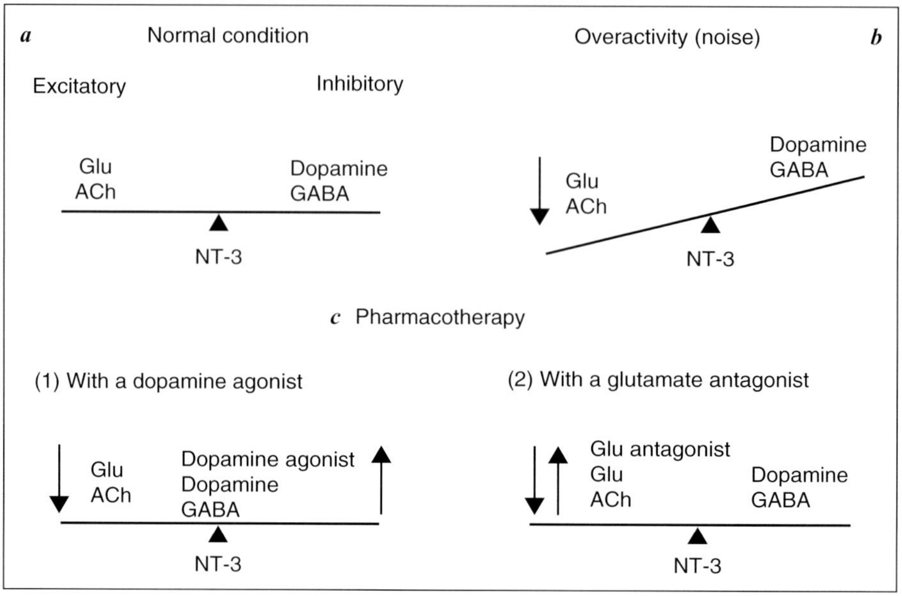

Fig. 4. Influence of *L*-glutamate, GABA, dopamine, acetylcholine (ACh) and NT-3 on the IHC synapse. ***a*** Balanced system under physiological conditions. ***b*** Possible situation under overexcitation. ***c*** Therapeutic strategy.

muscarinic receptor subtypes. The neurotrophic factor NT-3 is able to modulate the afferent neurotransmission presumably via NMDA/AMPA receptors. Thus a finely balanced situation between inhibitory and excitatory inputs seems to enable a gain control of the activity of afferent neurons of IHCs (fig. 4a). Considering this model, a situation of excitotoxicity like one under noise conditions would increase the glutamatergic input (fig. 4b). Several possible neuropharmacological strategies could be used to compensate this situation: adding a dopamine agonist, a GABA agonist or a glutamate antagonist (fig. 4b). The protective effects of dopamine agonists have already been studied [16]. Several studies have discussed the protective effects of glutamate antagonists. Antagonists of both iGluR showed protective properties in different animal models of inner ear diseases [17–21]. Especially a combination of synergistically acting drugs seems to increase the potency of a neuropharmacological treatment of inner ear diseases [21]. Although most glutamate antagonists investigated so far are not in clinical use, memantine (an NMDA antagonist) is one of the few drugs available for clinical use [22]. The severe side effects of glutamate antagonists in the central nervous system may be reduced by a local application to the inner ear [23].

References

1 Eybalin M: Neurotransmitter and neuromodulators in the mammalian cochlea. Physiol Rev 1993;73:309–373.

2 Pujol R, Puel JL, D'Aldain CG, Eybalin M: Pathophysiology of the glutamatergic synapses in the cochlea. Acta Otolaryngol (Stockh) 1993;113:330–334.

3 Lipton SA, Rosenberg PA: Excitatory amino acids as a final common pathway for neurologic disorders. N Engl J Med 1994;330:613–622.

4 Ehrenberger K, Felix D: Glutamate receptors in the afferent cochlear neurotransmission in guinea pigs. Hear Res 1991;52:73–80.

5 Ehrenberger K, Felix D: Receptor pharmacological models for inner ear therapies with emphasis on glutamate receptors: A survey. Acta Otolaryngol (Stockh) 1995;115:236–240.

6 Ruel J, Chen C, Pujol R, Bobbin RP, Puel JL: AMPA-preferring glutamate receptors in cochlear physiology of adult guinea-pig. J Physiol 1999;518:667–680.

7 Usami S, Matsubara A, Fujita S, Shinkawa H, Hayashi M: NMDA (NMDAR1) and AMPA-type (GluR2/3) receptor subunits are expressed in the inner ear. Neuroreport 1995;6:1161–1164.

8 Kleinlogel S, Oestreicher E, Arnold T, Ehrenberger K, Felix D: Metabotropic group I glutamate receptors modulate inner hair cell transmission. Neuroreport 1999;10:1879–1882.

9 Safieddine S, Eybalin M: Expression of mGluR1 alpha mRNA receptor in rat and guinea pig cochlear neurons. Neuroreport 1995;7:193–196.

10 Niedzielski AS, Safieddine S, Wenthold RJ: Molecular analysis of excitatory amino acid receptor expression in the cochlea. Audiol Neurootol 1997;2:79–91.

11 Oestreicher E, Arnold W, Ehrenberger K, Felix D: Dopamine regulates the glutamatergic inner hair cell activity in guinea pigs. Hear Res 1997;107:46–52.

12 Sun W, Salvi RJ: Dopamine modulates sodium currents in cochlear spiral ganglion neurons. Neuroreport 2001;12:803–807.

13 Arnold T, Oestreicher E, Ehrenberger K, Felix H, Felix D: GABA-A-receptor modulates the activity of inner hair cell afferents in guinea pig cochlea. Hear Res 1998;125:147–153.

14 Safieddine S, Bartolami S, Wenthold RJ, Eybalin M: Pre- and postsynaptic M_3 muscarinic receptor mRNAs in the rodent peripheral auditory system. Brain Res Mol Brain Res 1996;40:127–135.

15 Oestreicher E, Knipper M, Arnold A, Zenner HP, Felix D: Neurotrophin 3 potentiates glutamatergic responses of IHC afferents in the cochlea in vivo. Eur J Neurosci 2000;12:1584–1590.

16 d'Aldin C, Puel JL, Leducq R, Crambes O, Eybalin M, Pujol R: Effects of a dopaminergic agonist in the guinea pig cochlea. Hear Res 1995;90:202–211.

17 Puel JL, Pujol R, Tribillac F, Ladrech S, Eybalin M: Excitatory amino acid antagonists protect cochlear auditory neurons from excitotoxicity. J Comp Neurol 1994;341:241–256.

18 Ruel J, Bobbin RP, Vidal D, Pujol R, Puel JL: The selective AMPA receptor antagonist GYKI 53784 blocks action potential generation and excitotoxicity in the guinea pig cochlea. Neuropharmacology 2000;39:1959–1973.

19 Basile AS, Huang JM, Xie C, Webster D, Berlin C, Skolnick P: N-methyl-*D*-aspartate antagonists limit aminoglycoside antibiotic-induced hearing loss. Nat Med 1996;2:1338–1343.

20 Selvadurai DK, Etheridge S, Jones P, Mulheran M, Cook JA: Pharmacological protection of auditory function against noise and hypoxia with MK 801. Clin Otolaryngol 2000;25:570–576.

21 Duan M, Agerman K, Ernfors P, Canlon B: Complementary roles of neurotrophin 3 and a N-methyl-*D*-aspartate antagonist in the protection of noise and aminoglycoside-induced ototoxicity. Proc Natl Acad Sci USA 2000;97:7597–7602.

22 Oestreicher E, Arnold W, Ehrenberger K, Felix D: Memantine suppresses the glutamatergic neurotransmission of mammalian inner hair cells. ORL 1998;60:18–21.

23 Oestreicher E, Arnold W, Ehrenberger K, Felix D: New approaches for inner ear therapy with glutamate antagonists. Acta Otolaryngol (Stockh) 1999;119:174–178.

Dr. Elmar Oestreicher, ENT Department, Technical University of Munich,
Ismaninger Strasse 22, D–81675 Munich (Germany)
Tel. +49 89 41402370, Fax +49 89 41404853, E-Mail e.oestreicher@lrz.tum.de

Felix D, Oestreicher E (eds): Rational Pharmacotherapy of the Inner Ear.
Adv Otorhinolaryngol. Basel, Karger, 2002, vol 59, pp 140–148

Simulation of Methods for Drug Delivery to the Cochlear Fluids

Alec N. Salt

Department of Otolaryngology, Washington University School of Medicine, St. Louis, Mo., USA

Abstract

The inner ear fluids are remarkably 'unstirred' so that it cannot be assumed that applied drugs are dispersed throughout the fluid spaces. Calculation of the effective concentrations achieved when drugs are applied directly to the inner ear is made possible by simulations combining the physical processes involved in solute dispersal, which are diffusion, longitudinal fluid flow and clearances to other compartments. The approach has been validated in numerous experiments in which ion-selective electrodes were used to characterize the spread of marker substances in the cochlear fluids. The model incorporates the known size and geometry of cochlear fluid spaces for 6 species, including the guinea pig and the human. The simulator allows the dispersal of drugs or other substances to be approximated with knowledge of relatively few parameters. The simulation program is available on the Internet at http://oto.wustl.edu/cochlea/.

Direct methods to deliver drugs to the inner ear are increasingly being used in experimental studies and as a potential route for the clinical treatment of inner ear disorders. One approach is to apply a solution of the drug to the round window (RW) membrane via the middle ear space. The drug subsequently enters the cochlear perilymph without the potential problems associated with perforation of the otic capsule. In experimental animals, drugs have also been delivered in acute or chronic experiments by injecting them into the perilymphatic spaces through a catheter sealed into the scala tympani (ST). Although a number of variations of these techniques are used, the majority of studies have not attempted to quantify the actual drug levels achieved by their methods. It is only recently that attention has been focused on the levels of drugs achieved [1].

For a number of years, we have been performing experimental studies in which markers were applied to the fluid spaces by a variety of techniques and their spread was documented using ion-selective microelectrodes that were sealed into the cochlear fluid spaces [2–4]. From these studies, some of the most fundamental properties of the cochlear fluid spaces have been established. Early in the studies, we found that chemical markers did not spread along the length of the cochlea as rapidly as we had anticipated [2]. Our notion at the time that, due to the small size of the fluid spaces, substances would rapidly equilibrate, was found to be incorrect. In subsequent efforts to interpret the experimental findings, we made extensive use of theoretical calculations incorporating diffusion, longitudinal volume flow and other factors. The most flexible and complete analysis of data was provided by a model in which the cochlear fluid spaces were represented by numeric arrays where each element of the array corresponded to a 0.1-mm segment of the scala length. Associated with each element of the array was the cross-sectional area of the scala segment, permitting the specific dimensions and volumes of the cochlear fluid spaces to be incorporated. Effects of the multitude of processes contributing to solute movements, including e.g. diffusion, volume flow or clearance, were calculated for small time intervals (100 ms) and results were integrated over time. The algorithms on which calculations are based are given elsewhere [3]. Although the initial purpose of the model was to interpret experimental data, it became apparent that the simulations gave valuable insights into the issues associated with manipulations of the cochlear fluids and could have broader applications in the field. The simulator program was therefore rewritten with an easy-to-use graphic interface and named the 'Washington University Cochlear Fluids Simulator', as shown in figure 1. The current version of the program (version 1.4) is available for download from the Washington University website at http://oto.wustl.edu/cochlea/. While use of the simulator requires some understanding of the technical issues associated with cochlear fluid homeostasis, it is apparent that such an understanding is a necessary prerequisite for any interpretation of cochlear fluid manipulations. Our experimental and simulated findings clearly demonstrate that solute movements in the cochlear fluid spaces are complex, and no quantitative interpretation of drug or other solute levels is possible without considerations comparable to those presented below.

Processes Incorporated into Simulations

As the inner ear fluids do not circulate and are not otherwise mixed, it cannot be assumed that an applied drug or solute is quickly distributed throughout the entire ear. Therefore, it is not sufficient to know the total volume of the inner ear fluid spaces and the amount of drug applied in order to determine the

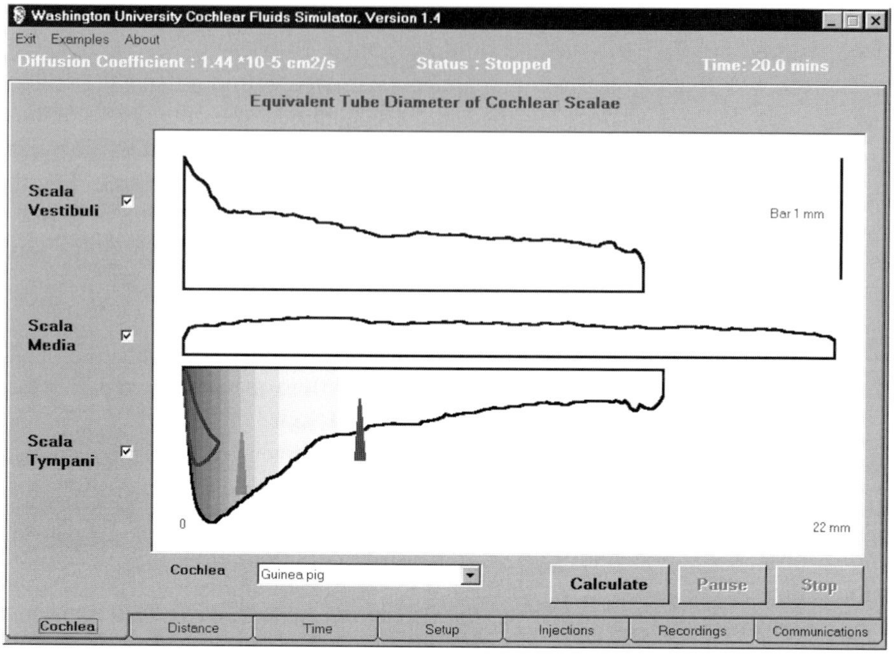

Fig. 1. Graphical display of the cochlear fluid spaces provided during simulation of solute entry into the ST of the guinea pig cochlea by application to the RW membrane. Scala dimensions were obtained by analysis of three-dimensional magnetic resonance images of guinea pig cochleae [5, 6].

concentration achieved by a specific manipulation. The major processes necessary to understand solute movements in the fluids of the ear, and incorporated into the simulator, are summarized in figure 2. They can be categorized as longitudinal and radial processes. Each specific solute does not necessarily require consideration of all the processes and one or more may be ignored for some solutes. The processes available in the simulator include those described below.

Dimensions and Anatomy of the Cochlear Fluid Spaces

The simulator takes into account scala lengths and cross-sectional areas as a function of distance for the cochleae of 6 species for which data are presently available. The species are the human, guinea pig, rat, bat, gerbil and mouse. Scala dimensions were derived in these species by analysis of three-dimensional images obtained by high-resolution magnetic resonance microscopy [5, 6]. Scala lengths were measured along the geometric midpoint of the scala. When measured in this manner, the scala media (SM) is substantially longer than the perilymphatic scalae. This is so because when seen in radial section (fig. 2, lower panel) the

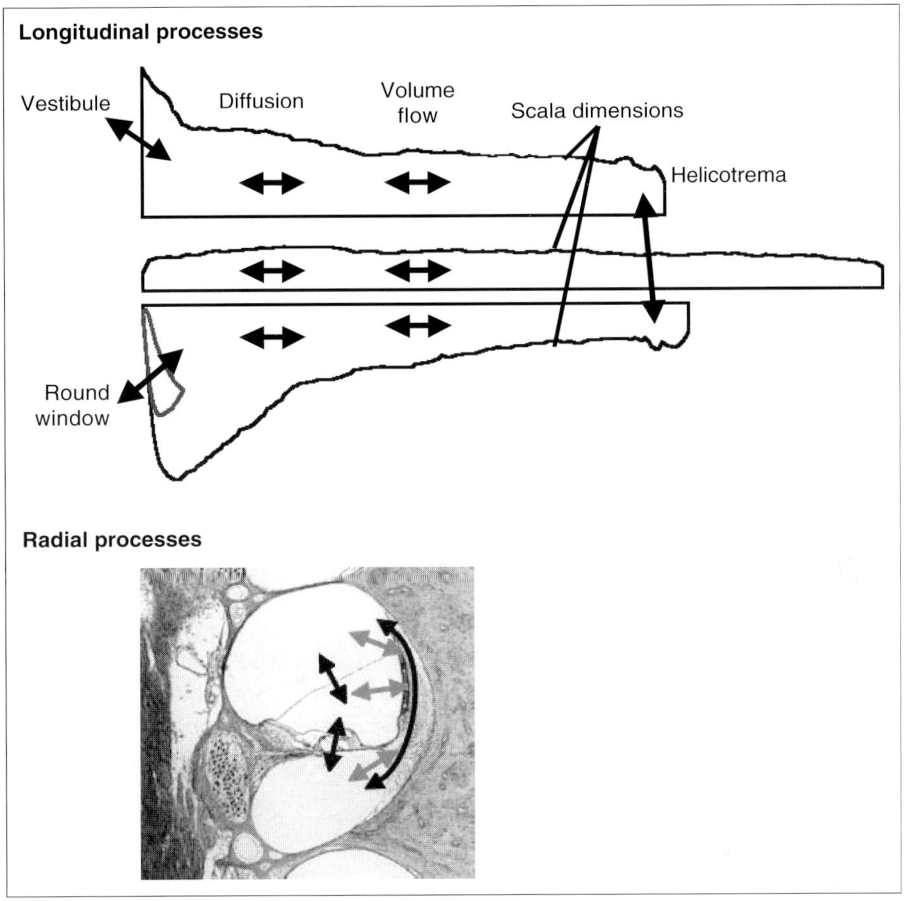

Fig. 2. Summary of the main solute dispersal processes incorporated into the simulator. The primary longitudinal factors include the scala dimensions, diffusion and volume flow as well as possible movements into the vestibule, across the helicotrema and across the RW membrane. In addition, radial processes are incorporated, including possible intercommunications with other scalae (black arrows) and clearances out of the fluid space (such as to blood or other spaces; gray arrows). In general, not all processes are involved in the dispersal of every substance so that some processes, such as longitudinal flow, can generally be ignored.

midpoint of the SM is lateral to that for both perilymphatic scalae, and as the SM follows a wider spiral around the modiolus, it is longer.

Diffusion

All solutes are influenced by diffusion, in which movements occur from regions of high concentration to regions of lower concentration. The rate of diffusion, represented by the diffusion coefficient, decreases as solute size increases [7]. If the diffusion coefficient for a solute has not been measured, it can be approximated on the basis of the molecular weight of the substance. In the cochlear fluids, diffusion appears to be the primary process by which solutes spread along the length of the fluid spaces. Even for small solutes, diffusion occurs relatively slowly for distances more than a few millimeters and may take many hours to spread over distances of more than 10 mm. Diffusion of larger molecules proceeds even more slowly. An appreciation of the nonlinear characteristics of diffusion is essential to understanding how drugs are dispersed in the cochlear fluids.

Longitudinal Flow

In the normal, sealed cochlea, rates of longitudinal endolymph and perilymph flow are exceedingly slow to the extent that their effects on solute dispersal can usually be neglected [8, 9]. In some instances, the incorporation of low flow rates can give a more precise fit to data recorded simultaneously in multiple cochlear locations, but the influence on the diffusion-generated concentration profiles along the scala is typically small. In contrast, when the cochlea is perforated, high rates of longitudinal flow occur through the perilymphatic spaces. In this state, flow rates of over 1 μl/min have been demonstrated in the guinea pig [8] and flow becomes the most significant process influencing solute movement. As this flow is caused by cerebrospinal fluid (CSF) entering the ST through the cochlear aqueduct, the washout of the perilymphatic space would be expected to have a profound influence on perilymphatic drug levels. For this reason, drug delivery to the ear usually requires the integrity of the otic capsule to be maintained.

Other Longitudinal Communications

The simulator takes into account the communication between the apical regions of the ST and scala vestibuli (SV) through the helicotrema as well as the spread from the basal end of the SV into the vestibule. In addition, the characteristics of drug entry through the RW membrane can be defined. For the guinea pig, RW area and the relationship of the RW to the fluid spaces at the base of the ST have been closely defined [10] and are incorporated into the simulator. In other species, these relationships can be approximated.

Interscalar Communications

Each radial segment of the scala has a potential communication with the other two scalae within the same segment, as shown in the lower part of figure 2. For example, the ST can communicate directly with both the SM and SV. The radial communication between the ST and SV has been demonstrated to be a relatively open communication route [3, 11] and is likely to be the main route by which vestibulotoxic drugs applied to the RW reach the vestibular system. One difficulty in quantifying these processes in terms of communication half-times arises from variations in scala area as a function of distance. Since identical homeostatic processes working into different volumes give different half-times, it is possible that the half-time of a process could vary along the length of the cochlea as the scala cross-sectional area changes. The documented cross-communication between the ST and SV was found to occur more rapidly in apical turns in a manner that was almost totally accounted for by scala area differences [3].

Clearance

Clearance is represented as a radial process occurring throughout the length of each scala. It encompasses all processes acting to reduce the level of the solute in the scala, including exchange with blood, spread into fluid spaces of the modiolus, into intracellular spaces and any binding or metabolism of the substance. In the simulator, clearance is quantified in terms of the effective half-time of the combined process. At present, it is assumed that the clearance half-time remains constant along the length of the scala.

Drug Application Methods

The simulation of drug application methods needs to take into account all the above fundamental processes affecting drug dispersal, in addition to simulation of the application process itself.

RW Application

Simulation of solute entry through the RW is established by a permeability coefficient and a defined RW area (determined anatomically). Experimental studies of entry of the ionic marker trimethylphenylammonium into the ST through the RW allowed kinetic parameters to be determined for the guinea pig cochlea [12]. In this study, RW permeability was found to be 19.4×10^{-9} cm/s, perilymph flow rate was apically directed at 4.4 nl/min, and ST clearance occurred with a half-time of 60 min. As with prior studies, perilymph flow rate was almost negligible. The calculated concentration profile along the ST at the end of a 90-min application of marker is shown in figure 3 (light line). The marker remains concentrated in the basal turn with minimal concentration reaching the apical turns. Based on the parameters established during 90-min application, the simulator was used to calculate the concentration profile after 24-hour application (fig. 3, heavy line). Even after prolonged application of the

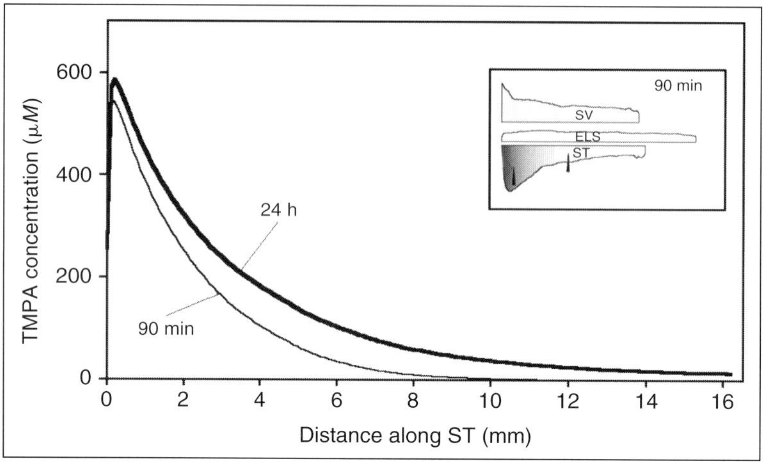

Fig. 3. Concentration profile of the marker trimethylphenylammonium (TMPA) along the ST after application of 2 m*M* solution to the RW. The profiles at 90 min (light line and inset image) are based on experimental recordings in the first and second turns of the ST. Using identical parameters, the concentration profile was calculated for 24-hour application (heavy line). A substantial base-to-apex concentration gradient remains even after 24 h. ELS = Endolymphatic space.

marker, a substantial base-to-apex concentration gradient remains along the scala. The explanation for this observation is that after a number of hours a steady state is established in which the marker is being cleared from the scala at a rate equal to the rate of diffusion. The normalized concentration profile is independent of the concentration in the middle ear and the RW permeability but depends on the diffusion rate and the rate of clearance of the substance from the ST. A uniform distribution of drug can only occur if there is no clearance of the drug from the cochlear fluid spaces. This study shows that one of the most important parameters establishing the drug level achieved in cochlear fluids is the rate at which the drug is cleared.

Injection into the Perilymph

Cannulae from miniosmotic pumps sealed into the basal turn of the ST permit drugs to be applied at controlled rates without the impediments and variability associated with passage across the RW membrane. Injections are typically performed at low rates (less than 0.01 μl/min) so that longitudinal flow associated with the injection is negligible. Under these conditions, the concentration profile along the length of the ST is very similar to that generated by drug application to the RW. Apical spread of drug along the ST again depends on the rate of solute diffusion, balanced by the rate of solute clearance from the scala. Even with higher injection rates (0.4 μl/min) where flow becomes a significant factor, experimental studies found results consistent with the flow being directed towards the cochlear aqueduct and therefore not contributing to the apical movement of drug along the ST [13]. A more even concentration profile can be achieved by making the injection site at a distance from the cochlear aqueduct. With injections into the cochlear apex at 0.4 μl/min, it was found that flow was again directed towards the cochlear aqueduct, resulting in a more uniform control over concentration in the ST. It is not yet known whether chronic injections into the cochlea at such high rates would be tolerated.

Perfusion

The established method of perilymphatic perfusion provides the most accurate control of drug level applied to the perilymphatic space. Since the cochlea is perforated, however, there is a rapid washout of the drug by CSF when perfusion ceases unless steps are taken to minimize CSF entry, by release of CSF pressure or by occlusion of the cochlear aqueduct. Perilymphatic perfusion is generally not appropriate for the delivery of drugs in chronic preparations.

Summary and Conclusions

With the use of simulations, we have been able to establish general principles underlying the spread of markers in cochlear fluids. Many parameters which are applicable to other solutes, including scala dimensions, flow rates and diffusion characteristics, have been derived. In order to simulate movements of a specific drug, experiments must be performed from which drug clearance can be derived. In addition, for drug application to the RW, the permeability properties of the RW membrane must be established. With these basic parameters, the characteristics of drug distribution can be closely approximated.

Basic principles of cochlear fluid manipulation include:

(1) substances applied to the intact cochlea do not rapidly disperse along the scalae; the degree of dispersal varies with species and cochlear size;

(2) longitudinal flow makes almost no contribution to solute movements in the intact cochlea; solute movements are dominated by passive diffusion;

(3) perforation of the otic capsule allows CSF to enter through the cochlear aqueduct, inducing flow from the base of the ST to the site of the outlet that rapidly displaces the native perilymph;

(4) injections into the sealed cochlea cause flow from the injection site towards the cochlear aqueduct which acts as an outlet. Perilymph replacement is only possible when the injection site is distant from the cochlear aqueduct.

Acknowledgement

This work was supported by research grant RO1 DC01368 from the National Institute on Deafness and Other Communication Disorders, National Institutes of Health.

References

1 Balough BJ, Hoffer ME, Wester D, O'Leary MJ, Brooker CR, Goto M: Kinetics of gentamycin uptake in the inner ear of chinchilla langier after middle ear administration in a sustained-release vehicle. Otolaryngol Head Neck Surg 1998;119:427–431.
2 Salt AN, Thalmann R, Marcus DC, Bohne BA: Direct measurement of longitudinal endolymph flow rate in the guinea pig cochlea. Hear Res 1986;23:141–151.
3 Salt AN, Ohyama K, Thalmann R: Radial communication between the perilymphatic scalae of the cochlea. 2. Estimation by bolus injection of tracer into the sealed cochlea. Hear Res 1991;56:37–43.
4 Salt AN, DeMott JE: Longitudinal endolymph flow associated with acute volume increase in the cochlea. Hear Res 1997;107:29–40.
5 Salt AN, Henson MM, Gewalt SL, Keating AW, De Mott JE, Henson OW: Detection and quantification of endolymphatic hydrops in the guinea pig cochlea by magnetic resonance microscopy. Hear Res 1995;88:79–86.
6 Thorne M, Salt AN, De Mott JE, Henson MM, Henson OW Jr, Gewalt SL: Cochlear fluid space dimensions for six species derived from reconstructions of 3-D magnetic resonance images. Laryngoscope 1999;109:1661–1668.
7 Hobbie RK: Intermediate Physics for Medicine and Biology. New York, Wiley, 1978.
8 Ohyama K, Salt AN, Thalmann R: Volume flow rate of perilymph in the guinea pig cochlea. Hear Res 1988;35:119–130.
9 Salt AN, Thalmann R: Rate of longitudinal flow of cochlear endolymph; in Nadol JB (ed): Ménière's Disease. Amsterdam, Kugler, 1989, pp 69–73.
10 Ghiz A, Salt AN, De Mott JE, Henson MM, Henson O: Quantification of inner ear structures from 3-D reconstruction of magnetic resonance images (abstract). Assoc Res Otolaryngol 2000;23:124.
11 Saijo S, Kimura RS: Distribution of HRP in the inner ear after injection into the middle ear cavity. Acta Otolaryngol (Stockh) 1984;97:593–610.
12 Salt AN, Ma Y: Quantification of solute entry into cochlear perilymph through the round window membrane. Hear Res 2001;154:88–97.
13 Ma Y, Salt AN: Effectiveness of manipulating cochlear perilymph by injections: A quantitative analysis (abstract). Assoc Res Otolaryngol 2000;23:125.

Dr. Alec N. Salt, Department of Otolaryngology,
Box 8115, Washington University School of Medicine,
660 South Euclid Avenue, St. Louis, MO 63110 (USA)
Tel. +1 314 362 7560, Fax +1 314 362 7522, E-Mail salta@msnotes.wustl.edu

Felix D, Oestreicher E (eds): Rational Pharmacotherapy of the Inner Ear.
Adv Otorhinolaryngol. Basel, Karger, 2002, vol 59, pp 149–155

Transtympanic Endoscopy for Drug Delivery to the Inner Ear Using a New Microendoscope

Stefan K.-R. Plontke[a], *Peter K. Plinkert*[a,b], *Beate Plinkert*[a,b], *Assen Koitschev*[a], *Hans-Peter Zenner*[a], *Hubert Löwenheim*[a]

Departments of Otorhinolaryngology, Head and Neck Surgery, University of [a] Tübingen and [b] Homburg/Saar, Germany

Abstract

Anatomic variations of the round window (RW) niche found in approximately 33% of human temporal bones may account for some of the problems associated with local drug delivery to the inner ear. A microendoscope with a total outer diameter of 1.2 mm was developed in particular for easy visualization and of drug delivery to the RW niche. It incorporated a thin fiber optic, a working/laser channel (0.3 mm) and an irrigation/suction channel (0.27 mm). When compared to a common 30° lens optic, with the microendoscope a greater area of the round window niche could be overseen. In addition, the endoscope could be advanced directly upon the surface of the RW membrane (RWM). The microendoscope may be used for evaluation of the anatomy of the RW niche prior to the placement of local drug delivery systems, for application of drugs directly onto the surface of the RWM or to verify the correct placement of inner ear drug delivery systems.

There has been an increasing interest in the treatment of inner ear disorders by local delivery of therapeutic agents directly onto the round window membrane (RWM). Conventional therapeutic strategies for inner ear disorders are systemic application of anti-inflammatory and rheologic drugs. However, treatment options are limited and some indications are even rather controversial. Moreover, because of poor control of drug level within the inner ear fluids, side effects or pharmacokinetic effects as, for instance, first-pass effects, many drugs may not be available for systemic use. Therefore, the idea of local inner ear therapy has been followed for some decades (for a review, see ref. [1]).

After the choice of the medication for the treatment of a specific inner ear disease, it is of importance that the drugs be appropriately delivered to their target, the inner ear. Drugs can be applied into the middle ear on the premise that they will diffuse through the RWM into the inner ear. Apart from our knowledge about the RWM structure [2], its permeability [3] and cochlear fluid flow [4], we have to consider anatomic variations of the RW niche.

A recent study on 202 temporal bones of 117 adult humans showed that the passage of medications to the RWM can be rendered more difficult by the presence of extraneous membranes stretched across the opening from the middle ear to the RW or by fibrous or fatty plugs obstructing the niche [5]. Extraneous RWMs are observed in 21.3% of the specimens. Fibrous plugs are found in 10.4% and fatty plugs in 1.5% of the cases. This means that in approximately 33% of the temporal bones the RWM is obstructed. The frequent obstruction of the RW niche is thought to explain the wide variations found in the dosage of medication required to produce a clinical result and treatment failures. To overcome this problem, it was recommended to examine the RW niche with endoscopes and to use any necessary surgical procedure to minimise the obstruction to the passage of the drugs to the inner ear [6, 7]. For this purpose, an endoscope which can be brought closely to the RWM in order to oversee its entire dimension is required.

Local drug delivery to the inner ear can be achieved by continuous and discontinuous drug delivery systems. There are different modalities to deliver a therapeutic agent to the surface of the RWM: transtympanic injection [8], application of sustained release vehicles [9], insertion of wicks through a transtympanic ventilation tube [7] or catheters placed in the RW niche and connected to external or implantable microdosage pumps [10–12].

Variations in the success rate and in the drug level needed for therapeutic results and treatment failures are not only attributable to the extraneous/false RWMs mentioned above but also due to incorrect placement or displacement of the device during treatment [12]. Therefore, the use of an endoscope is also recommended to control the correct placement at initial implantation and during treatment. The endoscopic procedure should be both minimally invasive and provide enough information to ensure correct placement of the medical device.

Finally, if the endoscope is fitted with a working channel and is small enough to reach the RW niche, small amounts of drugs can be delivered directly onto the RWM through a small myringotomy.

Methods

A microendoscope was modified in cooperation with the Explorent GmbH/Stuemed GmbH, Tuttlingen, Germany, in particular for easy visualization of the RW niche. It incorporated a thin 600-mm fiber optic with 6,000 pixels allowing a total endoscope diameter as small as

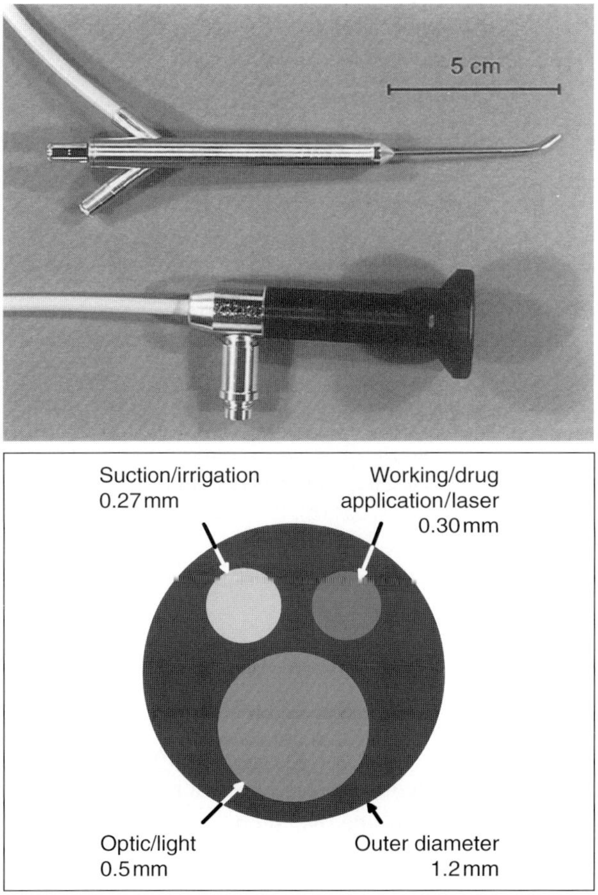

a
b

Fig. 1. Microendoscope developed in cooperation with Explorent GmbH/Stuemed GmbH® in particular for easy visualization of and drug delivery to the RW niche (***a***). It incorporates a thin fiber optic (6,000 pixels), a working/laser channel (0.3 mm i.d.) and an irrigation/suction channel (0.27 mm i.d.) allowing a total endoscope diameter as small as 1.2 mm. (***b***) Cross-sectional view.

1.2 mm (3.6 Fr.). The length of the endoscope was 50 mm. It included a working channel (inner diameter 0.3 mm, e.g. for delivery of substances or insertion of a laser fiber). It also has a suction irrigation channel with an inner diameter of 0.27 mm. The design features included a curved tip 5 mm in length with a 40° angle (fig. 1a, b).

The RW niches of 23 temporal bone preparations were approached through the ear channel with either a common 30° lens optic (outer diameter 2.7 mm and 1.9 mm; Karl Storz GmbH & Co., Tuttlingen, Germany) and with the new microendoscope. The two endoscopes were compared with regard to the visualization and accessibility of the RW niche and the quality of the image.

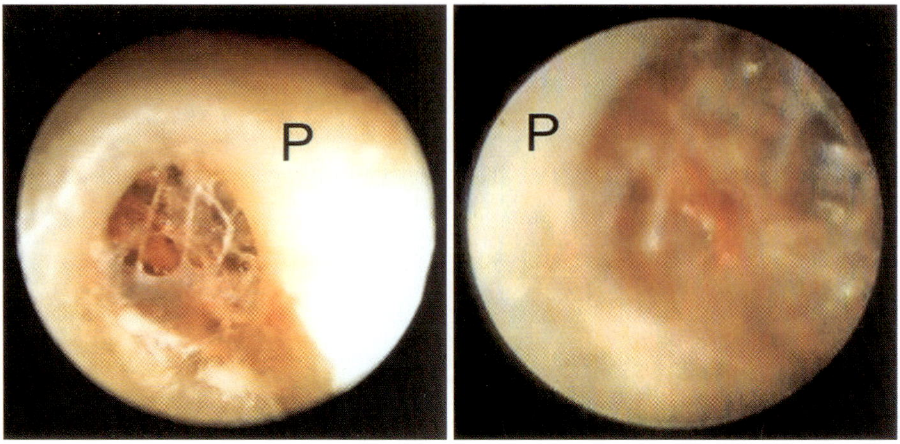

Fig. 2. RWM obstructed with false membranes in a temporal bone study as seen through a conventional lens optic (***a***) and the new microendoscope (***b***). P = Promontory.

Results

The evaluation of visualization and accessibility of the RW niche and quality of the image was done by a group of otologists on a subjective basis. Evaluation of the above criteria revealed that with this new microendoscope:

(1) A greater area of the RW niche could be overseen.
(2) The endoscope could be advanced directly to the surface of the RWM (fig. 2a, b).
(3) The correct placement of the two drug delivery systems – the RWμCath™ and the Micro-Wick™ – were successfully verified using the microendoscope (fig. 3a–c).
(4) The microendoscope created less trauma to the tympanic membrane.
(5) The image quality provided by the new microendoscope was sufficient to oversee and assess obstruction of the RW niche. However, the image quality of the common lens optic was clearly superior to that of the fiber optic (fig. 2a, b).

Discussion

Among the questions that need to be addressed to develop successful strategies of local inner ear therapy are (1) the medication that should be used,

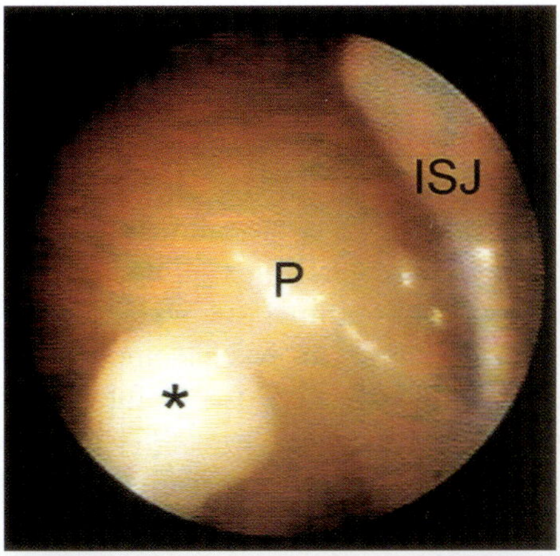

3a

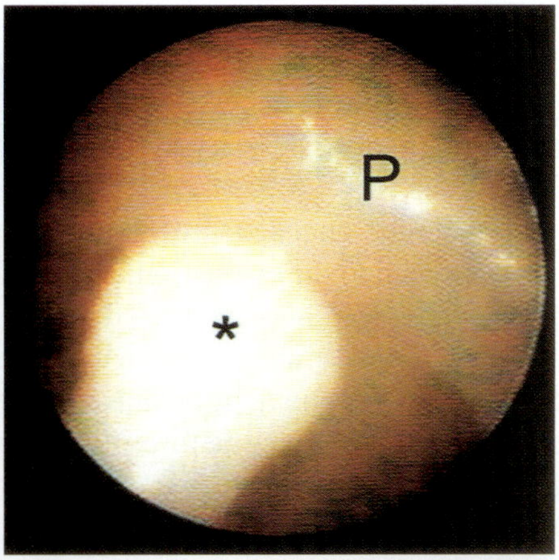

3b

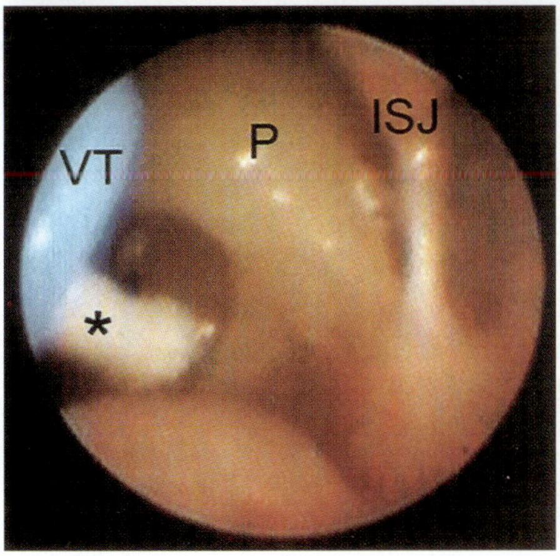

3c

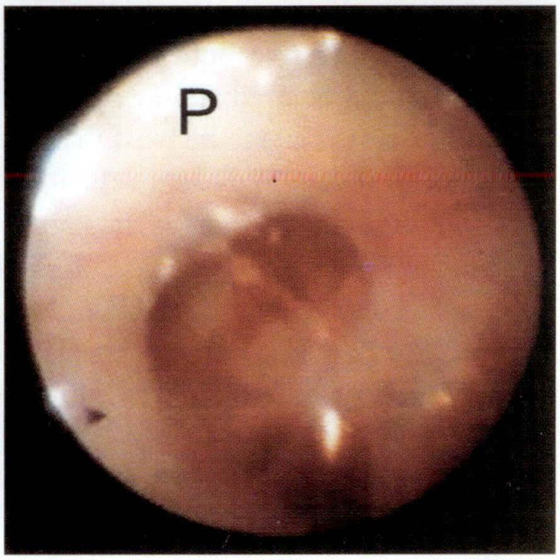

4

Fig. 3a–c. Correct placement of drug delivery systems (RWμCath™ and Micro-Wick™). ***a*** The ball tip (∗) of the RWμCath™ snugly fits into the RW niche. ***b*** Close up. ***c*** Tympanic side of the Micro-Wick™ (∗) in the RW niche in a temporal bone preparation. The image quality of the new microendoscope is sufficient for control of placement. P = Promontory; ISJ = incudostapedial joint; VT = ventilation tube.

Fig. 4. Intraoperative view onto the RWM using the new microendoscope. An unobstructed round window can be identified allowing the easy placement of a drug delivery system. P = Promontory.

(2) its best dose, concentration and preparation with regard to pharmacokinetics and toxicity and (3) the mode of delivery.

When drugs are applied via transtympanic injection with or without a vent tube [8, 13] or through a transtympanic Wick (e.g. Micro-Wick™) [7], this is done under the condition that no obstruction of the RW niche is interfering with the diffusion. Otherwise, treatment failure or unpredictable dosing might result. These obstructions occur singly or in combination in 33% of temporal bones [5]. Therefore it is necessary to carefully examine the RW niche before placement of any drug delivery system. This should be done with the least invasive procedure.

Although endoscopy of the middle ear is not new [14, 15], little attention has been paid to the evaluation of the RW niche. Since new drugs applied locally to the inner ear have appeared to be promising in animal studies and may undergo clinical trials in the near future [16–21], this issue has become a matter of interest. Some authors have described the use of an endoscope for evaluation of the RW niche before placement of a drug delivery system [7]. In our experience, however, when a common lens optic is used, even if its diameter is as small as 1.7 mm, the RW niche cannot be overseen completely in most cases due to the curvature of the outer ear canal or damage to the tympanic membrane is not acceptable. We believe that the new microendoscope with its smaller diameter and special shape is more suitable for gentle assessment of the RW niche prior to the placement of drug delivery systems. Patients most likely to benefit from this therapy by exclusion of obstruction can be selected.

Another important fact is displacement of drug delivery systems during therapy, which may account for some of the treatment failures [12]. The new microendoscope is small enough to allow a minimally invasive procedure to check for correct placement of a catheter in cases of treatment failures. The endoscope has now been manufactured in sterilizable form and is being clinically tested (fig. 4).

Conclusion

The new microendoscope proves as a helpful tool for visualization and direct access to the RW niche through a transtympanic approach. It may be used for evaluation of structures obstructing the RWM prior to the placement of drug delivery systems for local inner ear therapy and verification of correct placement of local drug delivery systems. The working channel of the microendoscope may be used for application of minute amounts of a drug directly onto its target, the surface of the RWM, through a minimally invasive procedure, i.e. a small myringotomy.

Acknowledgements

We thank Explorent GmbH/Stuemed GmbH and especially Mr. Charles Schraut for their cooperation and Mr. Kircher and Mr. Scheible of the University of Tübingen Center of information technology and Mr. Zurow for support in preparing the video prints.

References

1 Blakley BW: Clinical forum: A review of intratympanic therapy. Am J Otol 1997;18:520–531.
2 Schachern PA, Paparella MM, Duvall AJ 3d, Choo YB: The human round window membrane. An electron microscopic study. Arch Otolaryngol 1984;110:15–21.
3 Goycoolea MV, Lundman L: Round window membrane. Structure function and permeability: A review. Microsc Res Tech 1997;36:201–211.
4 Salt AN, DeMott JE: Longitudinal endolymph movements induced by perilymphatic injections. Hear Res 1998;123:137–147.
5 Alzamil KS, Linthicum FH Jr: Extraneous round window membranes and plugs: Possible effect on intratympanic therapy. Ann Otol Rhinol Laryngol 2000;109:30–32.
6 Silverstein H, Arruda J, Rosenberg SI, Deems D, Hester TO: Direct round window membrane application of gentamicin in the treatment of Ménière's disease. Otolaryngol Head Neck Surg 1999;120:649–655.
7 Silverstein H: Use of a new device, the MicroWick, to deliver medication to the inner ear. Ear Nose Throat J 1999;78:595–600.
8 Lange G: Technical difficulties in intratympanic gentamycin therapy in Ménière disease. Laryngol Rhinol Otol (Stuttg) 1985;64:174.
9 Hoffer ME, Balough B, Henderson J, DeCicco M, Wester D, O'Leary MJ, Kopke R: Use of sustained release vehicles in the treatment of Ménière's disease. Otolaryngol Clin North Am 1997; 30:1159–1166.
10 DeCicco MJ, Hoffer ME, Kopke RD, Wester D, Allen KA, Gottshall K, O'Leary MJ: Round-window microcatheter-administered microdose gentamicin: Results from treatment of tinnitus associated with Ménière's disease. Int Tinnitus J 1998;4:141–143.
11 Lehner R, Brugger H, Maassen MM, Zenner HP: A totally implantable drug delivery system for local therapy of the middle and inner ear. Ear Nose Throat J 1997;76:567–570.
12 Thomsen J, Charabi S, Tos M: Preliminary results of a new delivery system for gentamicin to the inner ear in patients with Ménière's disease. Eur Arch Otorhinolaryngol 2000;257:362–365.
13 Shea JJ Jr: The role of dexamethasone or streptomycin perfusion in the treatment of Ménière's disease. Otolaryngol Clin North Am 1997;30:1051–1059.
14 Thomassin JM: Otoendoscopically Guided Surgery. Paris, Springer, 1994, pp 1–87.
15 Mer SB, Derbyshire AJ, Brushenko BEE, Pontarelli DA: Fiberoptic endotoscopes for examining the middle ear. Arch Otolaryng 1967;85:61–67.
16 Oestreicher E, Arnold W, Ehrenberger K, Felix D: New approaches for inner ear therapy with glutamate antagonists. Acta Otolaryngol 1999;119:174–178.
17 Oestreicher E, Knipper M, Arnold A, Zenner HP, Felix D: Neurotrophin 3 potentiates glutamatergic responses of IHC afferents in the cochlea in vivo. Eur J Neurosci 2000;12:1584–1590.
18 Pirvola U, Xing-Qun L, Virkkala J, Saarma M, Murakata C, Camoratto AM, Walton KM, Ylikoski J: Rescue of hearing, auditory hair cells, and neurons by CEP-1347/KT7515, an inhibitor of c-Jun N-terminal kinase activation. J Neurosci 2001;20:43–50.
19 Stover T, Yagi M, Raphael Y: Cochlear gene transfer: Round window versus cochleostomy inoculation. Hear Res 1999;136:124–130.
20 Lalwani AK, Mhatre AN: Cochlear gene therapy. Adv Otorhinolaryngol 2000;56:275–278.
21 Löwenheim H, Furness DN, Kil J, Zinn C, Gultig K, Fero ML, Frost D, Gummer AW, Roberts JM, Rubel EW, Hackney CM, Zenner HP: Gene disruption of p27(Kip1) allows cell proliferation in the postnatal and adult organ of Corti. Proc Natl Acad Sci USA 1999;96:4084–4088.

Dr. med. Hubert Löwenheim, D/o ORL:HNS, University of Tübingen,
Silcherstrasse 5, D–72076 (Germany)
Tel. +49 70 71/29 84 411, Fax +49 70 71/29 33 11,
E-Mail Hubert.Loewenheim@uni-tuebingen.de

Felix D, Oestreicher E (eds): Rational Pharmacotherapy of the Inner Ear.
Adv Otorhinolaryngol. Basel, Karger, 2002, vol 59, pp 156–162

Clinical Experience with Caroverine in Inner Ear Diseases

Klaus Ehrenberger

Department of Otorhinolaryngology, University of Vienna, Austria

Abstract

The glutamatergic synapses between the cochlear inner hair cells and their afferent neurons seem to be mostly involved in the pathophysiology of the cochlea. Glutamatergic neurotoxicity is characterized by a mitochondrial overproduction of free oxygen radicals damaging lipid membranes and DNA structures of the postsynaptic neuron followed by the clinical symptoms of hearing loss and tinnitus. In preclinical tests, quinoxaline derivatives antagonized these deleterious consequences of too high an amount of free radicals. Therefore the clinically available quinoxaline dione caroverine provides a new approach to a successful treatment of tinnitus, sudden hearing loss and speech discrimination disorders in presbyacusis. The results of corresponding clinical trials are presented.

Because tinnitus, sudden hearing loss and progressive hearing loss are serious handicaps affecting millions of people in the world, any agent that can protect and/or regenerate the functional integrity of the auditory system would be highly beneficial. But, as the underlying causes of these cochlea-related symptoms are not clearly identified, a generally accepted treatment does not exist yet. Local and general circulatory disturbances, metabolic disorders, inflammatory reactions, mainly of viral etiology, noise and head traumas, and aging processes are emphasized to trigger these auditory impairments.

Considering the discussed multifactorial etiology of the symptoms, Ehrenberger and Felix [1] interpreted the origins of acquired hearing loss and tinnitus as evidence for a final common pathway of cochlear injury. In the mammalian cochlea, including that of the human, the excitatory amino acid glutamate mediates the neurotransmission between the inner hair cells and the afferent neurons [2–4]. Under different pathological conditions, the physiological transmitter glutamate exerts a uniform neurotoxic action [5, 6], and reactive oxygen species are incriminated to be the main mediators of tissue damage [for a review, see 7]. This suitable model of a glutamatergic otoneurotoxicity promises to make a crucial contribution to the understanding of the origin of a variety of inner ear diseases characterized by acute or progressive hearing loss and tinnitus [1].

In the cochlea of guinea pigs, quinoxaline derivatives prevent these pathological conditions [8] and trigger dendritic and synaptic repair mechanisms of glutamatergic units [9].

On the basis of these preclinical findings, we hypothesized that the clinical administration of quinoxaline derivatives would protect the human cochlea from neurotoxic attacks and stimulate intracochlear repair mechanisms necessary for the successful treatment of auditory symptoms.

To test this hypothesis, a study program was initiated using the quinoxaline derivative caroverine as specific test substance. Caroverine was chosen, because it is the only quinoxaline derivative successfully proven in preclinical phase I trials, is clinically available and is therefore free for the intended clinical test program.

Study Program Following Good Clinical Practice

Following the guidelines of the Center of Drug Evaluation and Research of the Federal Food and Drug Administration (FDA) of the USA, at the end of the preclinical phase I trial program, an early 'proof of concept' study is highly desirable to develop an understanding of the therapeutic potential of an agent in humans. Dose finding is the central challenge of phase II development. The subsequent phase III safety/efficacy trials finalize the whole study program which guarantees the best information on drug effectiveness.

In 1999, the European Federation for Pharmaceutical Sciences (EUFEPS) reconfirmed (6th EUFEPS Conference, Basel, Switzerland) the recommendations of the FDA and pointed out again that proof of concept trials represent the most important decision point to gain time and value in fast-tracked, informative drug development.

Remembering the above-mentioned considerations, studies conforming with Good Clinical Practice were designed and carried out separately for three different indications. The studies were performed according to the Declaration of Helsinki on Biomedical Research (Summerset West amendment), and were approved by the Ethics Commission of the University of Vienna Medical School. All subjects gave their written informed consent. The following statistical methods were used:

Study protocols were observed by intention-to-treat and per-protocol sets. Tabular statistical analyses of the data sets were done by StatXact-4, the Wilcoxon signed rank test for exact distributions and for Monte Carlo estimations of p values. Graphs of confidence intervals were computed by Student's t test. Additionally, for phase III, Fisher's exact test was used and descriptive statistics for safety parameters.

Results

Caroverine in the Treatment of Sudden Sensory Hearing Loss

Sudden sensory hearing loss represents an almost unilateral loss of hearing functions of more than 20 dB over at least three continuous audiometric frequencies occurring within 3 days or less. In a proof of concept study, we explored the action of caroverine in patients suffering from strictly unilateral severe sudden sensory hearing loss. They manifested no concomitant severe neurological, cardiovascular or metabolic disorders. With reference to the given spontaneous recovery rate, the neglect of which could create misinterpretations [10], only subjects with pure-tone hearing loss worse than 70 dB HL, having a generally unfavorable prognosis [11, 12] were included in this study.

The representative sample of patients was treated according to a standardized procedure for an exclusively intravenous administration of the quinoxaline derivative caroverine (Tinnex, Phafag AG, Schaanwald, Liechtenstein). Caroverine has been registered as a spasmolytic for a long time in Austria as well as in Switzerland. For this study, the preparation consisted of 250 ml of solution containing 160 mg of caroverine hydrochloride in 0.9% sodium chlorate for infusion, given twice a day with an infusion rate of 2 ml/min.

The acquired experimental and statistical data of the study suggest that caroverine, administered within the first 2 weeks after the onset of sudden sensory hearing loss, promises a highly significant therapeutic benefit in contrast to rheological and vasoactive drugs. In accordance with the recommendations of Mattox and Simmons [13], this caroverine-related effect can be classified as complete recovery for low and middle frequencies and as good recovery for high frequencies. These results offered the basis for the design of a subsequent multinational, multicenter phase II dose-finding study, which is still under way.

Caroverine in the Treatment of Age-Related Progressive Sensory Hearing Loss (Presbyacusis)

Presbyacusis is characterized by symmetric high-frequency hearing loss and impairment of speech discrimination. The development of presbyacusis has been described in numerous studies and the age-related annual threshold deteriorations were carefully noted [14]. Neural presbyacusis is thought to be a cumulative effect to repeated neurotoxic injuries of the peripheral endings of auditory nerve fibers [9, 15, 16]. Number and functional integrity of the auditory neurons determine the ability to discriminate speech [17].

On the other hand, there is an increasing body of evidence that age-related altered intracochlear blood flow and a pathological metabolism of reactive oxygen species result in hair cell damage [18]. In rats, long-term antioxidative management showed a protective effect on age-related hearing loss [18].

In a proof of concept study, the oral long-term administration of caroverine (60–180 mg/day, dependent on body weight) improved gradually the ability to recognize speech of a representative patient cohort suffering from presbyacusis. After 4–6 months of therapy, the evidence of this caroverine-induced effect was significant enough to continue the clinical test program and to design subsequent multicenter phase II and phase III studies.

In presbyacusis, pure-tone thresholds were not influenced by caroverine. Therefore, the cochlear hair-cell-linked amplifiers are caroverine insensitive. It can be concluded that the results of this proof of concept study support the hypothesis that caroverine triggers repair mechanisms intrinsic to the spiral ganglion neurons, the number and functional integrity of which decide the quality of sensorineural speech processing [17].

Caroverine in the Treatment of Tinnitus

Following the model of glutamatergic otoneurotoxicity, tinnitus is of cochlear-synaptic origin [1]. The validity of the model was tested in a proof of concept and phase II study program, which has recently been published [19].

A final multinational, multicenter phase III study was intended to show that a single infusion of caroverine, individually administered up to 160 mg, is more effective than placebo in the therapy of cochlear-synaptic tinnitus.

The design as well as the interpretation of the results of the phase III study must take into account a number of obstacles that are very specific to trials of prospective tinnitolytic agents. One factor is the ubiquitous nature of the complaint. Following an extensive international epidemiological study of Hiller et al. [20], initiated by the World Health Organization, tinnitus is clearly more frequent among patients with somatization disorders (42%) and/or hypochondriacal disorders (27%). Tinnitus is also related to depression, anxiety and to symptoms indicating autonomic arousal.

This frequent concurrence of tinnitus of sensorineural, organic etiology [1, 19, 21] and a psychopathological condition requires three different approaches to measure relevant tinnitus parameters:

(a) the organic dysfunction of the auditory system is assessed by 'objective' psychophysical tinnitus simulation tests (=tinnitus matching tests) and scaled in decibels;

(b) the tinnitus rating test is purely subjective in order to register introspectively the severity of the actual tinnitus sensation;

(c) the tinnitus-related condition is subjectively assessed as 'better', 'worse' or 'unchanged' after therapy.

It is for the first time that tinnitus intensity, tinnitus sensation and tinnitus-related condition were distinguished and analyzed separately in a clinical trial.

The described differentiated approaches to tinnitus phenomena elicited detailed and unexpected results concerning caroverine's efficacy. In 48% of the representative patient cohort (n = 70), the statistical evaluation of the data demonstrated a significant, stable, drug-induced reduction of the tinnitus intensity, at a minimum of 50% (in absolute values) in tinnitus simulation tests. In contrast, placebo elicited unstable responses decreasing in a distinct way during the first week after infusion. This evident therapeutic effect depends on a careful timing of the release of the adequate amount of caroverine. However, the subjective condition-related and condition-dependent tests showed different and protracted time courses. In 52% of the patients, caroverine did not significantly influence tinnitus. Therefore, the question arose, whether further tinnitus generators exist in the auditory system [22, 23].

Discussion

This is the first global study program on quinoxaline derivatives in clinical otology. The results of the trials demonstrate that the quinoxaline dione caroverine provides an improvement of sensorineural hearing loss and suppresses cochlear-synaptic tinnitus, both thought to be a consequence of glutamatergic neurotoxic attacks on afferent dendrites in the cochlea.

Some aspects of the presented caroverine studies merit discussion:

(1) The scientific proof of test results gains in significance crossing the clinical test program from proof of concept studies to phase III trial. Following the FDA guidelines for proof of concept, consideration should be given to the more modest goal of determining whether the pharmacological effect predicted from the preclinical development is present. The demonstrated evidence of a caroverine-induced hearing improvement in both proof of concept studies corresponds to this demand. Subsequent phase II and phase III studies for dose finding and placebo-controlled efficacy are still under way.

(2) The final phase III study testing the efficacy of caroverine in tinnitus treatment exhibited a complex interdependence of several parameters which are influenced by caroverine in a positive but different way. The study demonstrated that the suppressive action of caroverine on tinnitus intensity, evaluated by tinnitus simulation tests, is not imperatively accompanied by a simultaneous amelioration of the individual condition. This intrinsic dichotomy corresponds to a comparable phenomenon in patients suffering from chronic pain [24]. Caroverine is a potent suppressor of the activities of sensorineural organic tinnitus generators but not a psychopharmaceutical. Therefore, corresponding to the actual concept in the treatment of chronic pain, combining somatic pain 'killers' and antidepressive drugs, the tinnitus 'killer' caroverine should be part of an integral

psychophysical tinnitus treatment in order to relieve consistently tinnitus sensation and at the same time tinnitus condition.

(3) In tinnitus treatment, the therapeutic effect depends on a careful timing of the release of the adequate amount of caroverine. This process reflects a functional nonlinearity of tinnitus generators. In contrast, the successful procedure in hearing loss tests is related to a linear increase in the amount of caroverine. Therefore, it can be concluded that tinnitus and hearing loss are not different symptoms of the same pathophysiological process. The pathophysiological processes underlying both symptoms are certainly closely related but not identical.

In conclusion, the studies presented in this paper provide further evidence of the efficacy of caroverine in sensorineural hearing impairment and offer the drug for the treatment of cochlear-synaptic tinnitus [25]. At the same time, the results confirm the working hypothesis of glutamatergic otoneurotoxicity causing cochlear-synaptic tinnitus and some types of sensorineural hearing loss, and caroverine-linked otoneuroprotection and regeneration preventing and curing these auditory symptoms. In addition, future clinical studies should focus on caroverine-induced protection and regeneration of spiral ganglion neurons therefore optimizing the effectiveness of cochlear implants [26].

References

1 Ehrenberger K, Felix D: Receptor pharmacological model for inner ear therapies with emphasis on glutamate receptors: A survey. Acta Otolaryngol (Stockh) 1995;115:236–240.

2 Ehrenberger K, Felix D: Glutamate receptors in afferent cochlear neurotransmission in guinea pigs. Hear Res 1991;52:73–80.

3 Eybalin M: Neurotransmitters and neuromodulators of the mammalian cochlea. Physiol Rev 1993; 73:309–373.

4 Nordang L, Oestreicher E, Arnold W, Anniko M: Glutamate is the afferent neurotransmitter in the human cochlea. Acta Otolaryngol (Stockh) 2000;120:359–362.

5 Pujol R, Puel J-L, Gervais D'Aldin C, Eybalin M: Pathophysiology of the glutamatergic synapses in the cochlea. Acta Otolaryngol (Stockh) 1993;113:330–334.

6 Vasama JP, Linthicum FH Jr: Idiopathic sudden sensorineural hearing loss: Temporal bone histopathologic study. Ann Otol Rhinol Laryngol 2000;109:527–532.

7 Lubec G: The hydroxyl radical: From chemistry to human disease. J Invest Med 1996;44:324–346.

8 Ehrenberger K, Felix D: Caroverine depresses the activity of cochlear glutamate receptors in guinea pigs: In vivo model for drug-induced neuroprotection? Neuropharmacology 1992;31: 1259–1263.

9 Pujol R, Puel J-L: Excitotoxicity, synaptic repair, and functional recovery in the mammalian cochlea: A review of recent findings. Ann NY Acad Sci 1999;884:249–254.

10 Michel O, Jahns T, Joost-Enneking M, Neugebauer P, Streppel M, Stennert E: Das antiphlogistisch-rheologische Infusionsschema nach Stennert in der Behandlung von kochleovestibulären Störungen. HNO 2000;48:182–188.

11 Russolo M, Bianchi M: Prognosis and therapy of early acute idiopathic auditory failure. Audiology 1988;27:215–226.

12 Sano H, Okamoto M, Shitara T, Hirayama M: What kind of patients are suitable for evaluating the therapeutic effect of sudden deafness? Am J Otol 1998;19:579–583.

13 Mattox DE, Simmons FB: Natural history of sudden sensorineural hearing loss. Ann Otol Rhinol Laryngol 1977;86:463–480.
14 Jönsson R, Rosenhall U: Hearing in advanced age: A study in 85-, 88- and 90-year-old people. Audiology 1998;37:207–218.
15 Lipsitz LA, Goldberger AL: Loss of 'complexity' and aging. JAMA 1992;267:1806–1809.
16 Felix H, Gleeson MJ, Pollak A, Johnsson L-G: The cochlear neurons in humans; in Iurato S, Veldman JE (eds): Progress in Human Auditory and Vestibular Histopathology. Amsterdam, Kugler, 1997, pp 73–79.
17 Svozil K, Felix D, Ehrenberger K: Multiple-channel fractal information coding of mammalian nerve signals. Biochem Biophys Res Commun 1994;199:911–915.
18 Seidman M: Effects of dietary restriction and antioxidants on presbyacusis. Laryngoscope 2000;110:727–738.
19 Denk D-M, Heinzl H, Franz P, Ehrenberger K: Caroverine in tinnitus treatment: A placebo-controlled blind study. Acta Otolaryngol (Stockh) 1997;117:825–830.
20 Hiller W, Janca A, Burke KC: Association between tinnitus and somatoform disorders. J Psychosom Res 1997;43:613–624.
21 Eggermont JJ: On the pathophysiology of tinnitus: A review and a peripheral model. Hear Res 1990;48:111–124.
22 Zenner HP, Ernst A: Cochlear-motor, transduction and signal-transfer tinnitus: Models for three types of cochlear tinnitus. Eur Arch Otorhinolaryngol 1993;249:447–454.
23 Lockwood AH, Salvi RJ, Coad ML, Towsley ML, Wack DS, Murphy BW: The functional neuroanatomy of tinnitus: Evidence for limbic system links and neural plasticity. Neurology 1998; 50:114–120.
24 Tonndorf J: The analogy between tinnitus and pain: A suggestion for a physiological basis of chronic tinnitus. Hear Res 1987;28:271–275.
25 Simpson JJ, Davies WE: Recent advances in the pharmacological treatment of tinnitus. Trends Pharmacol Sci 1999;20:12–18.
26 Marzella PL, Clark GM: Growth factors, auditory neurons and cochlear implants: A review. Acta Otolaryngol (Stockh) 1999;119:407–412.

Prof. Klaus Ehrenberger, MD, Department of Otorhinolaryngology of the University of Vienna, General Hospital, Währinger Gürtel 18–20, A–1097 Vienna (Austria)
Tel. +43 1 40 400 3305, Fax +43 1 405 01 704, E-Mail HNO@UNIVIE.AC.AT

Author Index

Subject Index